T0351058

Graduate Texts in Mathematics 12

Graduate Texts in Mathematics

continued after Index

Richard Beals

Advanced Mathematical Analysis

Periodic Functions and Distributions,
Complex Analysis, Laplace Transform
and Applications

Springer-Verlag
New York Berlin Heidelberg
London Paris Tokyo

Richard Beals
Department of Mathematics
Yale University
New Haven, CT 06520
U.S.A.

AMS Subject Classification
46-01, 46S05, 46C05, 30-01, 43-01
34-01, 3501

Library of Congress Cataloging in Publication Data

Beals, Richard, 1938-
    Advanced mathematical analysis.

    (Graduate texts in mathematics, v. 12)
    1. Mathematical analysis. I. Title. II. Series.
QA300.B4   515   73-6884

Third Corrected Printing, 1987.

ISBN 0-387-90065-9 Springer-Verlag New York Heidelberg Berlin (hard cover)
ISBN 3-540-90065-9 Springer-Verlag Berlin Heidelberg New York (hard cover)

to Nancy

PREFACE

Once upon a time students of mathematics and students of science or engineering took the same courses in mathematical analysis beyond calculus. Now it is common to separate "advanced mathematics for science and engineering" from what might be called "advanced mathematical analysis for mathematicians." It seems to me both useful and timely to attempt a reconciliation.

The separation between kinds of courses has unhealthy effects. Mathematics students reverse the historical development of analysis, learning the unifying abstractions first and the examples later (if ever). Science students learn the examples as taught generations ago, missing modern insights. A choice between encountering Fourier series as a minor instance of the representation theory of Banach algebras, and encountering Fourier series in isolation and developed in an *ad hoc* manner, is no choice at all.

It is easy to recognize these problems, but less easy to counter the legitimate pressures which have led to a separation. Modern mathematics has broadened our perspectives by abstraction and bold generalization, while developing techniques which can treat classical theories in a definitive way. On the other hand, the applier of mathematics has continued to need a variety of definite tools and has not had the time to acquire the broadest and most definitive grasp—to learn necessary and sufficient conditions when simple sufficient conditions will serve, or to learn the general framework encompassing different examples.

This book is based on two premises. First, the ideas and methods of the theory of distributions lead to formulations of classical theories which are satisfying and complete mathematically, and which at the same time provide the most useful viewpoint for applications. Second, mathematics and science students alike can profit from an approach which treats the particular in a careful, complete, and modern way, and which treats the general as obtained by abstraction for the purpose of illuminating the basic structure exemplified in the particular. As an example, the basic L^2 theory of Fourier series can be established quickly and with no mention of measure theory once $L^2(0, 2\pi)$ is known to be complete. Here $L^2(0, 2\pi)$ is viewed as a subspace of the space of periodic distributions and is shown to be a Hilbert space. This leads to a discussion of abstract Hilbert space and orthogonal expansions. It is easy to derive necessary and sufficient conditions that a formal trigonometric series be the Fourier series of a distribution, an L^2 distribution, or a smooth function. This in turn facilitates a discussion of smooth solutions and distribution solutions of the wave and heat equations.

The book is organized as follows. The first two chapters provide background material which many readers may profitably skim or skip. Chapters 3, 4, and 5 treat periodic functions and distributions, Fourier series, and applications. Included are convolution and approximation (including the

Weierstrass theorems), characterization of periodic distributions, elements of Hilbert space theory, and the classical problems of mathematical physics. The basic theory of functions of a complex variable is taken up in Chapter 6. Chapter 7 treats the Laplace transform from a distribution-theoretic point of view and includes applications to ordinary differential equations. Chapters 6 and 7 are virtually independent of the preceding three chapters; a quick reading of sections 2, 3, and 5 of Chapter 3 may help motivate the procedure of Chapter 7.

I am indebted to Max Jodeit and Paul Sally for lively discussions of what and how analysts should learn, to Nancy for her support throughout, and particularly to Fred Flowers for his excellent handling of the manuscript.

Added for second printing: I am very grateful to several colleagues, in particular to Ronald Larsen and to S. Dierolf, for their lists of errors.

Added for third printing: Robert Burckel provided an extensive list of errors and comments, for which I am also grateful.

Richard Beals

TABLE OF CONTENTS

Chapter Five   Applications of Fourier series

Chapter Six   Complex analysis

Chapter Seven   The Laplace transform

Advanced Mathematical Analysis

Chapter 1

Basic Concepts

§1. Sets and functions

One feature of modern mathematics is the use of abstract concepts to provide a language and a unifying framework for theories encompassing numerous special cases and examples. Two important examples of such concepts, that of "metric space" and that of "vector space," will be taken up later in this chapter. In this section we discuss briefly the concepts, even more basic, of "set" and of "function."

We assume that the intuitive notion of a "set" and of an "element" of a set are familiar. A set is determined when its elements are specified in some manner. The exact manner of specification is irrelevant, provided the elements are the same. Thus

$$A = \{3, 5, 7\}$$

means that A is the set with three elements, the integers 3, 5, and 7. This is the same as

$$A = \{7, 3, 5\},$$

or

$$A = \{n \mid n \text{ is an odd positive integer between 2 and 8}\}$$

or

$$A = \{2n + 1 \mid n = 1, 2, 3\}.$$

In expressions such as the last two, the phrase after the vertical line is supposed to prescribe exactly what precedes the vertical line, thus prescribing the set. It is convenient to allow repetitions; thus A above is also

$$\{5, 3, 7, 3, 3\},$$

still a set with three elements. If x is an element of A we write

$$x \in A \quad \text{or} \quad A \ni x.$$

If x is not an element of A we write

$$x \notin A \quad \text{or} \quad A \not\ni x.$$

The sets of all integers and of all positive integers are denoted by $\mathbb{Z}$ and $\mathbb{Z}_+$ respectively:

$$\mathbb{Z} = \{0, 1, -1, 2, -2, 3, -3, \ldots\},$$
$$\mathbb{Z}_+ = \{1, 2, 3, 4, \ldots\}.$$

As usual the three dots ... indicate a presumed understanding about what is omitted.

1

Other matters of notation:

$\varnothing$ denotes the *empty set* (no elements).

$A \cup B$ denotes the *union*, $\{x \mid x \in A \text{ or } x \in B \text{ (or both)}\}$.

$A \cap B$ denotes the *intersection*, $\{x \mid x \in A \text{ and } x \in B\}$.

The union of $A_1, A_2, \ldots, A_m$ is denoted by

$$A_1 \cup A_2 \cup A_3 \cup \cdots \cup A_m \quad \text{or} \quad \bigcup_{j=1}^{m} A_j,$$

and the intersection by

$$A_1 \cap A_2 \cap A_3 \cap \cdots \cap A_m \quad \text{or} \quad \bigcap_{j=1}^{m} A_j.$$

The union and the intersection of an infinite family of sets $A_1, A_2 \ldots$ indexed by $\mathbb{Z}_+$ are denoted by

$$\bigcup_{j=1}^{\infty} A_j \quad \text{and} \quad \bigcap_{j=1}^{\infty} A_j.$$

More generally, suppose J is a set, and suppose that for each $j \in J$ we are given a set A_j. The union and intersection of all the A_j are denoted by

$$\bigcup_{j \in J} A_j \quad \text{and} \quad \bigcap_{j \in J} A_j.$$

A set A is a *subset* of a set B if every element of A is an element of B; we write

$$A \subset B \quad \text{or} \quad B \supset A.$$

In particular, for any A we have $\varnothing \subset A$. If $A \subset B$, the *complement of A in B* is the set of elements of B not in A:

$$B \backslash A = \{x \mid x \in B, x \notin A\}.$$

Thus $C = B \backslash A$ is equivalent to the two conditions

$$A \cup C = B, \qquad A \cap C = \varnothing.$$

The *product* of two sets A and B is the set of ordered pairs (x, y) where $x \in A$ and $y \in B$; this is written $A \times B$. More generally, if $A_1, A_2, \ldots, A_n$ are the sets then

$$A_1 \times A_2 \times \cdots \times A_n$$

is the set whose elements are all the ordered n-tuples $(x_1, x_2, \ldots, x_n)$, where each $x_j \in A_j$. The product

$$A \times A \times \cdots \times A$$

of n copies of A is also written A^n.

A *function* from a set A to a set B is an assignment, to each element of A, of some unique element of B. We write

$$f \colon A \to B$$

for a function f from A to B. If $x \in A$, then $f(x)$ denotes the element of B assigned by f to the element x. The elements assigned by f are often called *values*. Thus a *real-valued function on A* is a function $f: A \to \mathbb{R}$, $\mathbb{R}$ the set of real numbers. A *complex-valued function on A* is a function $f: A \to \mathbb{C}$, $\mathbb{C}$ the set of complex numbers.

A function $f: A \to B$ is said to be *1-1* ("one-to-one") or *injective* if it assigns distinct elements of B to distinct elements of A: If $x, y \in A$ and $x \neq y$, then $f(x) \neq f(y)$. A function $f: A \to B$ is said to be *onto* or *surjective* if for each element $y \in B$, there is some $x \in A$ such that $f(x) = y$. A function $f: A \to B$ which is both 1-1 and onto is said to be *bijective*.

If $f: A \to B$ and $g: B \to C$, the *composition of f and g* is the function denoted by $g \circ f$:

$$g \circ f: A \to C, \qquad g \circ f(x) = g(f(x)), \qquad \text{for all } x \in A.$$

If $f: A \to B$ is bijective, there is a unique *inverse* function $f^{-1}: B \to A$ with the properties: $f^{-1} \circ f(x) = x$, for all $x \in A$; $f \circ f^{-1}(y) = y$, for all $y \in B$.

Examples

Consider the functions $f: \mathbb{Z} \to \mathbb{Z}_+$, $g: \mathbb{Z} \to \mathbb{Z}$, $h: \mathbb{Z} \to \mathbb{Z}$, defined by

$$\begin{aligned} f(n) &= n^2 + 1, &\quad n \in \mathbb{Z}, \\ g(n) &= 2n, &\quad n \in \mathbb{Z}, \\ h(n) &= 1 - n, &\quad n \in \mathbb{Z}. \end{aligned}$$

Then f is neither 1-1 nor onto, g is 1-1 but not onto, h is bijective, $h^{-1}(n) = 1 - n$, and $f \circ h(n) = n^2 - 2n + 2$.

A set A is said to be *finite* if either $A = \varnothing$ or there is an $n \in \mathbb{Z}_+$, and a bijective function f from A to the set $\{1, 2, \ldots, n\}$. The set A is said to be *countable* if there is a bijective $f: A \to \mathbb{Z}_+$. This is equivalent to requiring that there be a bijective $g: \mathbb{Z}_+ \to A$ (since if such an f exists, we can take $g = f^{-1}$; if such a g exists, take $f = g^{-1}$). The following elementary criterion is convenient.

Proposition 1.1. *If there is a surjective (onto) function $f: \mathbb{Z}_+ \to A$, then A is either finite or countable.*

Proof. Suppose A is not finite. Define $g: \mathbb{Z}_+ \to \mathbb{Z}_+$ as follows. Let $g(1) = 1$. Since A is not finite, $A \neq \{f(1)\}$. Let $g(2)$ be the first integer m such that $f(m) \neq f(1)$. Having defined $g(1), g(2), \ldots, g(n)$, let $g(n + 1)$ be the first integer m such that $f(m) \notin \{f(1), f(2), \ldots, f(n)\}$. The function g defined inductively on all of $\mathbb{Z}_+$ in this way has the property that $f \circ g: \mathbb{Z}_+ \to A$ is bijective. In fact, it is 1-1 by the construction. It is onto because f is onto and by the construction, for each n the set $\{f(1), f(2), \ldots, f(n)\}$ is a subset of $\{f \circ g(1), f \circ g(2), \ldots, f \circ g(n)\}$. ☐

Corollary 1.2. *If B is countable and $A \subset B$, then A is finite or countable.*

Proof. If $A = \varnothing$, we are done. Otherwise, choose a function $f: \mathbb{Z}_+ \to B$ which is onto. Choose an element $x_0 \in A$. Define $g: \mathbb{Z}_+ \to A$ by: $g(n) = f(n)$ if $f(n) \in A$, $g(n) = x_0$ if $f(n) \notin A$. Then g is onto, so A is finite or countable. □

Proposition 1.3. *If* $A_1, A_2, A_3, \ldots$ *are finite or countable, then the sets*

$$\bigcup_{j=1}^{n} A_j \quad and \quad \bigcup_{j=1}^{\infty} A_j$$

are finite or countable.

Proof. We shall prove only the second statement. If any of the A_j are empty, we may exclude them and renumber. Consider only the second case. For each A_j we can choose a surjective function $f_j: \mathbb{Z}_+ \to A_j$. Define $f: \mathbb{Z}_+ \to \bigcup_{j=1}^{\infty} A_j$ by $f(1) = f_1(1)$, $f(3) = f_1(2)$, $f(5) = f_1(3), \ldots, f(2) = f_2(1)$, $f(6) = f_2(2)$, $f(10) = f_2(3), \ldots$, and in general $f(2^{j-1}(2k - 1)) = f_j(k)$, $j, k = 1, 2, 3, \ldots$. Any $x \in \bigcup_{j=1}^{\infty} A_j$ is in some A_j, and therefore there is $k \in \mathbb{Z}_+$ such that $f_j(k) = x$. Then $f(2^{j-1}(2k - 1)) = x$, so f is onto. By Proposition 1.1, $\bigcup_{j=1}^{\infty} A_j$ is finite or countable. □

Example

Let $\mathbb{Q}$ be the set of rational numbers: $\mathbb{Q} = \{m/n \mid m \in \mathbb{Z}, n \in \mathbb{Z}_+\}$. This is countable. In fact, let $A_n = \{j/n \mid j \in \mathbb{Z}, -n^2 \le j \le n^2\}$. Then each A_n is finite, and $\mathbb{Q} = \bigcup_{n=1}^{\infty} A_n$.

Proposition 1.4. *If* $A_1, A_2, \ldots, A_n$ *are countable sets, then the product set* $A_1 \times A_2 \times \cdots \times A_n$ *is countable.*

Proof. Choose bijective functions $f_j: A_j \to \mathbb{Z}_+$, $j = 1, 2, \ldots, n$. For each $m \in \mathbb{Z}_+$, let B_m be the subset of the product set consisting of all n-tuples $(x_1, x_2, \ldots, x_n)$ such that each $f_j(x_j) \le m$. Then B_m is finite (it has m^n elements) and the product set is the union of the sets B_m. Proposition 1.3 gives the desired conclusion. □

A *sequence* in a set A is a collection of elements of A, not necessarily distinct, indexed by some countable set J. Usually J is taken to be $\mathbb{Z}_+$ or $\mathbb{Z}_+ \cup \{0\}$, and we use the notations

$$(a_n)_{n=1}^{\infty} = (a_1, a_2, a_3, \ldots),$$
$$(a_n)_{n=0}^{\infty} = (a_0, a_1, a_2, \ldots).$$

Proposition 1.5. *The set S of all sequences in the set $\{0, 1\}$ is neither finite nor countable.*

Proof. Suppose $f: \mathbb{Z}_+ \to S$. We shall show that f is not surjective. For each $m \in \mathbb{Z}_+$, $f(m)$ is a sequence $(a_{n,m})_{n=1}^{\infty} = (a_{1,m}, a_{2,m}, \ldots)$, where each $a_{n,m}$ is 0 or 1. Define a sequence $(a_n)_{n=1}^{\infty}$ by setting $a_n = 0$ if $a_{n,n} = 1$, $a_n = 1$ if $a_{n,n} = 0$. Then for each $m \in \mathbb{Z}_+$, $(a_n)_{n=1}^{\infty} \ne (a_{n,m})_{n=1}^{\infty} = f(m)$. Thus f is not surjective. □

We introduce some more items of notation. The symbol $\Rightarrow$ means "implies"; the symbol $\Leftarrow$ means "is implied by"; the symbol $\Leftrightarrow$ means "is equivalent to."

Anticipating §2 somewhat, we introduce the notation for *intervals* in the set $\mathbb{R}$ of real numbers. If $a, b \in \mathbb{R}$ and $a < b$, then

$$(a, b) = \{x \mid x \in \mathbb{R}, a < x < b\},$$
$$(a, b] = \{x \mid x \in \mathbb{R}, a < x \leq b\},$$
$$[a, b) = \{x \mid x \in \mathbb{R}, a \leq x < b\},$$
$$[a, b] = \{x \mid x \in \mathbb{R}, a \leq x \leq b\}.$$

Also,

$$(a, \infty) = \{x \mid x \in \mathbb{R}, a < x\},$$
$$(-\infty, a] = \{x \mid x \in \mathbb{R}, x \leq a\}, \quad \text{etc.}$$

§2. Real and complex numbers

We denote by $\mathbb{R}$ the set of all real numbers. The operations of addition and multiplication can be thought of as functions from the product set $\mathbb{R} \times \mathbb{R}$ to $\mathbb{R}$. Addition assigns to the ordered pair (x, y) an element of $\mathbb{R}$ denoted by $x + y$; multiplication assigns an element of $\mathbb{R}$ denoted by xy. The algebraic properties of these functions are familiar.

Axioms of addition

A1. $(x + y) + z = x + (y + z)$, for any $x, y, z \in \mathbb{R}$.
A2. $x + y = y + x$, for any $x, y \in \mathbb{R}$.
A3. There is an element 0 in $\mathbb{R}$ such that $x + 0 = x$ for every $x \in \mathbb{R}$.
A4. For each $x \in \mathbb{R}$ there is an element $-x \in \mathbb{R}$ such that $x + (-x) = 0$.

Note that the element 0 is unique. In fact, if $0'$ is an element such that $x + 0' = x$ for every x, then

$$0' = 0' + 0 = 0 + 0' = 0.$$

Also, given x the element $-x$ is unique. In fact, if $x + y = 0$, then

$$y = y + 0 = y + (x + (-x)) = (y + x) + (-x)$$
$$= (x + y) + (-x) = 0 + (-x) = (-x) + 0 = -x.$$

This uniqueness implies $-(-x) = x$, since $(-x) + x = x + (-x) = 0$.

Axioms of multiplication

M1. $(xy)z = x(yz)$, for any $x, y, z \in \mathbb{R}$.
M2. $xy = yx$, for any $x, y \in \mathbb{R}$.
M3. There is an element $1 \neq 0$ in $\mathbb{R}$ such that $x1 = x$ for any $x \in \mathbb{R}$.
M4. For each $x \in \mathbb{R}$, $x \neq 0$, there is an element x^{-1} in $\mathbb{R}$ such that $xx^{-1} = 1$.

Note that 1 and x^{-1} are unique. We leave the proofs as an exercise.

Distributive law

DL. $x(y + z) = xy + xz$, for any $x, y, z \in \mathbb{R}$.

Note that DL and A2 imply $(x + y)z = xz + yz$.
We can now readily deduce some other well-known facts. For example,

$$0 \cdot x = (0 + 0) \cdot x = 0 \cdot x + 0 \cdot x,$$

so $0 \cdot x = 0$. Then

$$x + (-1) \cdot x = 1 \cdot x + (-1) \cdot x = (1 + (-1)) \cdot x = 0 \cdot x = 0,$$

so $(-1) \cdot x = -x$. Also,

$$(-x) \cdot y = ((-1) \cdot x) \cdot y = (-1) \cdot (xy) = -xy.$$

The axioms A1–A4, M1–M4, and DL do not determine $\mathbb{R}$. In fact there is a set consisting of two elements, together with operations of addition and multiplication, such that the axioms above are all satisfied: if we denote the elements of the set by 0, 1, we can define addition and multiplication by

$$0 + 0 = 1 + 1 = 0, \qquad 0 + 1 = 1 + 0 = 1,$$
$$0 \cdot 0 = 1 \cdot 0 = 0 \cdot 1 = 0, \qquad 1 \cdot 1 = 1.$$

There is an additional familiar notion in $\mathbb{R}$, that of *positivity*, from which one can derive the notion of an ordering of $\mathbb{R}$. We axiomatize this by introducing a subset $P \subset \mathbb{R}$, the set of "positive" elements.

Axioms of order

O1. If $x \in \mathbb{R}$, then exactly one of the following holds: $x \in P$, $x = 0$, or $-x \in P$.

O2. If $x, y \in P$, then $x + y \in P$.

O3. If $x, y \in P$, then $xy \in P$.

It follows from these that if $x \neq 0$, then $x^2 \in P$. In fact if $x \in P$ then this follows from O3, while if $-x \in P$, then $(-x)^2 \in P$, and $(-x)^2 = -(x(-x)) -(-x^2) = x^2$. In particular, $1 = 1^2 \in P$.

We *define* $x < y$ if $y - x \in P$, $x > y$ if $y < x$. It follows that $x \in P \Leftrightarrow x > 0$. Also, if $x < y$ and $y < z$, then

$$z - x = (z - y) + (y - x) \in P,$$

so $x < z$. In terms of this order, we introduce the *Archimedean axiom*.

O4. If $x, y > 0$, then there is a positive integer n such that $nx = x + x + \cdots + x$ is $> y$.

(One can think of this as saying that, given enough time, one can empty a large bathtub with a small spoon.)

The axioms given so far still do not determine $\mathbb{R}$; they are all satisfied by the subset $\mathbb{Q}$ of rational numbers. The following notions will make a distinction between these two sets.

A nonempty subset $A \subset \mathbb{R}$ is said to be *bounded above* if there is an $x \in \mathbb{R}$ such that every $y \in A$ satisfies $y \le x$ (as usual, $y \le x$ means $y < x$ or $y = x$). Such a number x is called an *upper bound* for A. Similarly, if there is an $x \in \mathbb{R}$ such that every $y \in A$ satisfies $x \le y$, then A is said to be *bounded below* and x is called a *lower bound* for A.

A number $x \in \mathbb{R}$ is said to be a *least upper bound* for a nonempty set $A \subset \mathbb{R}$ if x is an upper bound, and if every other upper bound x' satisfies $x' \ge x$. If such an x exists it is clearly unique, and we write

$$x = \text{lub } A.$$

Similarly, x is a *greatest lower bound* for A if it is a lower bound and if every other lower bound x' satisfies $x' \le x$. Such an x is unique, and we write

$$x = \text{glb } A.$$

The final axiom for $\mathbb{R}$ is called the *completeness axiom*.

O5. If A is a nonempty subset of $\mathbb{R}$ which is bounded above, then A has a least upper bound.

Note that if $A \subset \mathbb{R}$ is bounded below, then the set $B = \{x \mid x \in \mathbb{R}, -x \in A\}$ is bounded above. If $x = \text{lub } B$, then $-x = \text{glb } A$. Therefore O5 is equivalent to: a nonempty subset of $\mathbb{R}$ which is bounded below has a greatest lower bound.

Theorem 2.1. $\mathbb{Q}$ *does not satisfy the completeness axiom.*

Proof. Recall that there is no rational p/q, $p, q \in \mathbb{Z}$, such that $(p/q)^2 = 2$: in fact if there were, we could reduce to lowest terms and assume either p or q is odd. But $p^2 = 2q^2$ is even, so p is even, so $p = 2m$, $m \in \mathbb{Z}$. Then $4m^2 = 2q^2$, so $q^2 = 2m^2$ is even and q is also even, a contradiction.

Let $A = \{x \mid x \in \mathbb{Q}, x^2 < 2\}$. This is nonempty, since $0, 1 \in A$. It is bounded above, since $x \ge 2$ implies $x^2 \ge 4$, so 2 is an upper bound. We shall show that no $x \in \mathbb{Q}$ is a least upper bound for A.

If $x \le 0$, then $x < 1 \in A$, so x is not an upper bound. Suppose $x > 0$ and $x^2 < 2$. Suppose $h \in \mathbb{Q}$ and $0 < h < 1$. Then $x + h \in \mathbb{Q}$ and $x + h > x$. Also, $(x + h)^2 = x^2 + 2xh + h^2 < x^2 + 2xh + h = x^2 + (2x + 1)h$. If we choose $h > 0$ so small that $h < 1$ and $h < (2 - x^2)/(2x + 1)$, then $(x + h)^2 < 2$. Then $x + h \in A$, and $x + h > x$, so x is not an upper bound of A.

Finally, suppose $x \in \mathbb{Q}$, $x > 0$, and $x^2 > 2$. Suppose $h \in \mathbb{Q}$ and $0 < h < x$. Then $x - h \in \mathbb{Q}$ and $x - h > 0$. Also, $(x - h)^2 = x^2 - 2xh + h^2 > x^2 - 2xh$. If we choose $h > 0$ so small that $h < 1$ and $h < (x^2 - 2)/2x$, then $(x - h)^2 > 2$. It follows that if $y \in A$, then $y < x - h$. Thus $x - h$ is an upper bound for A less than x, and x is not the least upper bound. $\square$

We used the non-existence of a square root of 2 in $\mathbb{Q}$ to show that O5 does not hold. We may turn the argument around to show, using O5, that there is

a real number $x > 0$ such that $x^2 = 2$. In fact, let $A = \{y \mid y \in \mathbb{R}, y^2 < 2\}$.
The argument proving Theorem 2.1 proves the following: A is bounded
above; its least upper bound x is positive; if $x^2 < 2$ then x would not be an
upper bound, while if $x^2 > 2$ then x would not be the least upper bound.
Thus $x^2 = 2$.

Two important questions arise concerning the above axioms. Are the
axioms consistent, and satisfied by some set $\mathbb{R}$? Is the set of real numbers the
only set satisfying these axioms?

The consistency of the axioms and the existence of $\mathbb{R}$ can be demonstrated
(to the satisfaction of most mathematicians) by *constructing* $\mathbb{R}$, starting with
the rationals.

In one sense the axioms do not determine $\mathbb{R}$ uniquely. For example, let
$\mathbb{R}^0$ be the set of all symbols x^0, where x is (the symbol for) a real number.
Define addition and multiplication of elements of $\mathbb{R}^0$ by

$$x^0 + y^0 = (x + y)^0, \quad x^0 y^0 = (xy)^0.$$

Define P^0 by $x^0 \in P^0 \Leftrightarrow x \in P$. Then $\mathbb{R}^0$ satisfies the axioms above. This is
clearly fraudulent: $\mathbb{R}^0$ is just a copy of $\mathbb{R}$. It can be shown that any set with
addition, multiplication, and a subset of positive elements, which satisfies all
the axioms above, is just a copy of $\mathbb{R}$.

Starting from $\mathbb{R}$ we can construct the set $\mathbb{C}$ of complex numbers, without
simply postulating the existence of a "quantity" i such that $i^2 = -1$. Let $\mathbb{C}^0$
be the product set $\mathbb{R}^2 = \mathbb{R} \times \mathbb{R}$, whose elements are ordered pairs (x, y) of
real numbers. Define addition and multiplication by

$$(x, y) + (x', y') = (x + x', y + y'),$$
$$(x, y)(x', y') = (xx' - yy', xy' + x'y).$$

It can be shown by straightforward calculations that $\mathbb{C}^0$ together with these
operations satisfies A1, A2, M1, M2, and DL. To verify the remaining
algebraic axioms, note that

$$(x, y) + (0, 0) = (x, y).$$
$$(x, y) + (-x, -y) = (0, 0),$$
$$(x, y)(1, 0) = (x, y),$$
$$(x, y)(x/(x^2 + y^2), -y/(x^2 + y^2)) = (1, 0) \qquad \text{if } (x, y) \neq (0, 0).$$

If $x \in \mathbb{R}$, let x^0 denote the element $(x, 0) \in \mathbb{C}^0$. Let i^0 denote the element
$(0, 1)$. Then we have

$$(x, y) = (x, 0) + (0, y) = (x, 0) + (0, 1)(y, 0) = x^0 + i^0 y^0.$$

Also, $(i^0)^2 = (0, 1)(0, 1) = (-1, 0) = -1^0$. Thus we can write any element
of $\mathbb{C}^0$ uniquely as $x^0 + i^0 y^0$, $x, y \in \mathbb{R}$, where $(i^0)^2 = -1^0$. We now drop the
superscripts and write $x + iy$ for $x^0 + i^0 y^0$ and $\mathbb{C}$ for $\mathbb{C}^0$: this is legitimate,
since for elements of $\mathbb{R}$ the new operations coincide with the old: $x^0 + y^0 = (x + y)^0$, $x^0 y^0 = (xy)^0$. Often we shall denote elements of $\mathbb{C}$ by z or w. When

we write $z = x + iy$, we shall understand that x, y are real. They are called the *real part* and the *imaginary part* of z, respectively:

$$z = x + iy, \qquad x = \text{Re}\,(z), \qquad y = \text{Im}\,(z).$$

There is a very useful operation in $\mathbb{C}$, called *complex conjugation*, defined by:

$$z^* = (x + iy)^* = x - iy.$$

Then z^* is called the *complex conjugate* of z. It is readily checked that

$$(z + w)^* = z^* + w^*, \qquad (zw)^* = z^*w^*,$$
$$(z^*)^* = z, \qquad z^*z = x^2 + y^2.$$

Thus $z^*z \neq 0$ if $z \neq 0$. Define the *modulus* of z, $|z|$, by

$$|z| = (z^*z)^{1/2} = (x^2 + y^2)^{1/2}, \qquad z = x + iy.$$

Then if $z \neq 0$,

$$1 = z^*z|z|^{-2} = z(z^*|z|^{-2}),$$

or

$$z^{-1} = z^*|z|^{-2}.$$

Adding and subtracting gives

$$z + z^* = 2x, \qquad z - z^* = 2iy \qquad \text{if } z = x + iy.$$

Thus

$$\text{Re}\,(z) = \tfrac{1}{2}(z + z^*), \qquad \text{Im}\,(z) = \tfrac{1}{2}i^{-1}(z - z^*).$$

The usual geometric representation of $\mathbb{C}$ is by a coordinatized plane: $z = x + iy$ is represented by the point with coordinates (x, y). Then by the Pythagorean theorem, $|z|$ is the distance from (the point representing) z to the origin. More generally, $|z - w|$ is the distance from z to w.

Exercises

1. There is a unique real number $x > 0$ such that $x^3 = 2$.
2. Show that $\text{Re}\,(z + w) = \text{Re}\,(z) + \text{Re}\,(w)$, $\text{Im}\,(z + w) = \text{Im}\,(z) + \text{Im}\,(w)$.
3. Suppose $z = x + iy$, $x, y \in \mathbb{R}$. Then

$$|x| \leq |z|, \qquad |y| \leq |z|, \qquad |z| \leq |x| + |y|.$$

4. For any $z, w \in \mathbb{C}$,

$$|zw^*| = |z|\,|w|.$$

5. For any $z, w \in \mathbb{C}$,

$$|z + w| \leq |z| + |w|.$$

(Hint: $|z + w|^2 = (z + w)^*(z + w) = |z|^2 + 2 \operatorname{Re}(zw^*) + |w|^2$; apply Exercises 3 and 4 to estimate $|\operatorname{Re}(zw^*)|$.)

6. The Archimedean axiom O4 can be deduced from the other axioms for the real numbers. (Hint: use O5).

7. If $a > 0$ and n is a positive integer, there is a unique $b > 0$ such that $b^n = a$.

§3. Sequences of real and complex numbers

A sequence $(z_n)_{n=1}^{\infty}$ of complex numbers is said to *converge to* $z \in \mathbb{C}$ if for each $\varepsilon > 0$, there is an integer N such that $|z_n - z| < \varepsilon$ whenever $n \geq N$. Geometrically, this says that for *any* circle with center z, the numbers z_n all lie inside the circle, except for possibly finitely many values of n. If this is the case we write

$$z_n \to z, \quad \text{or} \quad \lim_{n \to \infty} z_n = z, \quad \text{or} \quad \lim z_n = z.$$

The number z is called the *limit* of the sequence $(z_n)_{n=1}^{\infty}$. Note that the limit is unique: suppose $z_n \to z$ and also $z_n \to w$. Given any $\varepsilon > 0$, we can take n so large that $|z_n - z| < \varepsilon$ and also $|z_n - w| < \varepsilon$. Then

$$|z - w| \leq |z - z_n| + |z_n - w| < \varepsilon + \varepsilon = 2\varepsilon.$$

Since this is true for all $\varepsilon > 0$, necessarily $z = w$.

The following proposition collects some convenient facts about convergence.

Proposition 3.1.   *Suppose $(z_n)_{n=1}^{\infty}$ and $(w_n)_{n=1}^{\infty}$ are sequences in $\mathbb{C}$.*

(a) $z_n \to z$ *if and only if* $z_n - z \to 0$.

(b) *Let* $z_n = x_n + iy_n$, x_n, y_n *real. Then* $z_n \to z = x + iy$ *if and only if* $x_n \to x$ *and* $y_n \to y$.

(c) *If* $z_n \to z$ *and* $w_n \to w$, *then* $z_n + w_n \to z + w$.

(d) *If* $z_n \to z$ *and* $w_n \to w$, *then* $z_n w_n \to zw$.

(e) *If* $z_n \to z \neq 0$, *then there is an integer M such that* $z_n \neq 0$ *if* $n \geq M$.

Moreover $(z_n^{-1})_{n=M}^{\infty}$ *converges to* z^{-1}.

Proof. (a) This follows directly from the definition of convergence.

(b) By Exercise 3 of §2,

$$2|x_n - x| + |y_n - y| \leq |z_n - z| \leq 2|x_n - x| + 2|y_n - y|.$$

It follows easily that $z_n - z \to 0$ if and only if $x_n - x \to 0$ and $y_n - y \to 0$.

(c) This follows easily from the inequality

$$|(z_n + w_n) - (z + w)| = |(z_n - z) + (w_n - w)| \leq |z_n - z| + |w_n - w|.$$

(d) Choose M so large that if $n \geq M$, then $|z_n - z| < 1$. Then for $n \geq M$,

$$|z_n| = |(z_n - z) + z| < 1 + |z|.$$

Let $K = 1 + |w| + |z|$. Then for all $n \geq M$,

$$
\begin{aligned}
|z_n w_n - zw| &= |z_n(w_n - w) + (z_n - z)w| \\
&\leq |z_n| \, |w_n - w| + |z_n - z| \, |w| \\
&\leq K(|w_n - w| + |z_n - z|).
\end{aligned}
$$

Since $w_n - w \to 0$ and $z_n - z \to 0$, it follows that $z_n w_n - zw \to 0$.

   (e) Take M so large that $|z_n - z| \leq \frac{1}{2}|z|$ when $n \geq M$. Then for $n \geq M$,

$$
\begin{aligned}
|z_n| &= |z_n| + \tfrac{1}{2}|z| - \tfrac{1}{2}|z| \\
&\geq |z_n| + |z - z_n| - \tfrac{1}{2}|z| \geq |z_n + (z - z_n)| - \tfrac{1}{2}|z| = \tfrac{1}{2}|z|.
\end{aligned}
$$

Therefore, $z_n \neq 0$. Also for $n \geq M$.

$$
\begin{aligned}
|z_n{}^{-1} - z^{-1}| &= |(z - z_n)z^{-1}z_n{}^{-1}| \\
&\leq |z - z_n| \cdot |z|^{-1} \cdot (\tfrac{1}{2}|z|)^{-1} = K|z - z_n|,
\end{aligned}
$$

where $K = 2|z|^{-2}$. Since $z - z_n \to 0$ we have $z_n{}^{-1} - z^{-1} \to 0$. □

   A sequence $(z_n)_{n=1}^{\infty}$ in $\mathbb{C}$ is said to be *bounded* if there is an $M \geq 0$ such that $|z_n| \leq M$ for all n; in other words, there is a fixed circle around the origin which encloses all the z_n's.

   A sequence $(x_n)_{n=1}^{\infty}$ in $\mathbb{R}$ is said to be *increasing* if for each n, $x_n \leq x_{n+1}$; it is said to be *decreasing* if for each n, $x_n \geq x_{n+1}$.

   Proposition 3.2. *A bounded, increasing sequence in $\mathbb{R}$ converges. A bounded, decreasing sequence in $\mathbb{R}$ converges.*

   Proof. Suppose $(x_n)_{n=1}^{\infty}$ is a bounded, increasing sequence. Then the set $\{x_n \mid n = 1, 2, \ldots\}$ is bounded above. Let x be its least upper bound. Given $\varepsilon > 0$, $x - \varepsilon$ is not an upper bound, so there is an N such that $x_N \geq x - \varepsilon$. If $n \geq N$, then

$$
x - \varepsilon \leq x_N \leq x_n \leq x,
$$

so $|x_n - x| \leq \varepsilon$. Thus $x_n \to x$. The proof for a decreasing sequence is similar. □

   If $A \subset \mathbb{R}$ is bounded above, the least upper bound of A is often called the *supremum* of A, written sup A. Thus

$$
\sup A = \operatorname{lub} A.
$$

Similarly, the greatest lower bound of a set $B \subset \mathbb{R}$ which is bounded below is also called the *infimum* of A, written inf A:

$$
\inf A = \operatorname{glb} A.
$$

   Suppose $(x_n)_{n=1}^{\infty}$ is a bounded sequence of reals. We shall associate with this given sequence two other sequences, one increasing and the other decreasing. For each n, let $A_n = \{x_n, x_{n+1}, x_{n+2}, \ldots\}$, and set

$$
x_n' = \inf A_n, \qquad x_n'' = \sup A_n.
$$

Now $A_n \supset A_{n+1}$, so any lower or upper bound for A_n is a lower or upper bound for A_{n+1}. Thus

$$x_n' \leq x_{n+1}', \qquad x_{n+1}'' \leq x_n''.$$

Choose M so that $|x_n| \leq M$, all n. Then $-M$ is a lower bound and M an upper bound for each A_n. Thus

(3.1) $$-M \leq x_n' \leq x_n'' \leq M, \qquad \text{all } n.$$

We may apply Proposition 3.2 to the bounded increasing sequence $(x_n')_{n=1}^\infty$ and the bounded decreasing sequence $(x_n'')_{n=1}^\infty$ and conclude that both converge. We define

$$\liminf x_n = \lim x_n',$$
$$\limsup x_n = \lim x_n''.$$

These numbers are called the *lower limit* and the *upper limit* of the sequence $(x_n)_{n=1}^\infty$, respectively. It follows from (3.1) that

(3.2) $$-M \leq \liminf x_n \leq \limsup x_n \leq M.$$

A sequence $(z_n)_{n=1}^\infty$ in $\mathbb{C}$ is said to be a *Cauchy sequence* if for each $\varepsilon > 0$ there is an integer N such that $|z_n - z_m| < \varepsilon$ whenever $n \geq N$ and $m \geq N$. The following theorem is of fundamental importance.

Theorem 3.3. *A sequence in $\mathbb{C}$ (or $\mathbb{R}$) converges if and only if it is a Cauchy sequence.*

Proof. Suppose first that $z_n \to z$. Given $\varepsilon > 0$, we can choose N so that $|z_n - z| \leq \frac{1}{2}\varepsilon$ if $n \geq N$. Then if $n, m \geq N$ we have

$$|z_n - z_m| \leq |z_n - z| + |z - z_m| < \tfrac{1}{2}\varepsilon + \tfrac{1}{2}\varepsilon = \varepsilon.$$

Conversely, suppose $(z_n)_{n=1}^\infty$ is a Cauchy sequence. We consider first the case of a real sequence $(x_n)_{n=1}^\infty$ which is a Cauchy sequence. The sequence $(x_n)_{n=1}^\infty$ is bounded: in fact, choose M so that $|x_n - x_m| < 1$ if $n, m \geq M$. Then if $n \geq M$,

$$|x_n| \leq |x_n - x_M| + |x_M| < 1 + |x_M|.$$

Let $K = \max\{|x_1|, |x_2|, \ldots, |x_{M-1}|, |x_M| + 1\}$. Then for any n, $|x_n| \leq K$. Now since the sequence is bounded, we can associate the sequences $(x_n')_{n=1}^\infty$ and $(x_n'')_{n=1}^\infty$ as above. Given $\varepsilon > 0$, choose N so that $|x_n - x_m| < \varepsilon$ if $n, m \geq N$. Now suppose $n \geq m \geq N$. It follows that

$$x_m - \varepsilon \leq x_n \leq x_m + \varepsilon, \qquad n \geq m \geq N.$$

By definition of x_n' we also have, therefore,

$$x_m - \varepsilon \leq x_n' \leq x_m + \varepsilon, \qquad n \geq m \geq N.$$

Letting $x = \liminf x_n = \lim x_n'$, we have

$$x_m - \varepsilon \leq x \leq x_m + \varepsilon, \qquad m \geq N,$$

or $|x_m - x| \leq \varepsilon$, $m \geq N$. Thus $x_n \to x$.

Now consider the case of a complex Cauchy sequence $(z_n)_{n=1}^{\infty}$. Let $z_n = x_n + iy_n$, x_n, $y_n \varepsilon \mathbb{R}$. Since $|x_n - x_m| \leq |z_n - z_m|$, $(x_n)_{n=1}^{\infty}$ is a Cauchy sequence. Therefore $x_n \to x \in \mathbb{R}$. Similarly, $y_n \to y \in \mathbb{R}$. By Proposition 3.1(b), $z_n \to x + iy$. □

The importance of this theorem lies partly in the fact that it gives a criterion for the *existence* of a limit in terms of the sequence itself. An immediately recognizable example is the sequence

$$3, \quad 3.1, \quad 3.14, \quad 3.142, \quad 3.1416, \quad 3.14159, \ldots,$$

where successive terms are to be computed (in principle) in some specified way. This sequence can be shown to be a Cauchy sequence, so we know it has a limit. Knowing this, we are free to give the limit a name, such as "π".

We conclude this section with a useful characterization of the upper and lower limits of a bounded sequence.

Proposition 3.4. *Suppose $(x_n)_{n=1}^{\infty}$ is a bounded sequence in $\mathbb{R}$. Then* $\liminf x_n$ *is the unique number x' such that*

(i)' *for any $\varepsilon > 0$, there is an N such that $x_n > x' - \varepsilon$ whenever $n \geq N$,*
(ii)' *for any $\varepsilon > 0$ and any N, there is an $n \geq N$ such that $x_n < x' + \varepsilon$.*

Similarly, $\limsup x_n$ *is the unique number x'' with the properties*

(i)" *for any $\varepsilon > 0$, there is an N such that $x_n < x'' + \varepsilon$ whenever $n \geq N$,*
(ii)" *for any $\varepsilon > 0$ and any N, there is an $n \geq N$ such that $x_n > x - \varepsilon$.*

Proof. We shall prove only the assertion about $\liminf x_n$. First, let $x_n' = \inf\{x_n, x_{n+1}, \ldots\} = \inf A_n$ as above, and let $x' = \lim x_n' = \liminf x_n$. Suppose $\varepsilon > 0$. Choose N so that $x_N' > x' - \varepsilon$. Then $n \geq N$ implies $x_n \geq x_N' > x' - \varepsilon$, so (i)' holds. Given $\varepsilon > 0$ and N, we have $x_N' \leq x' < x' + \frac{1}{2}\varepsilon$. Therefore $x' + \frac{1}{2}\varepsilon$ is not a lower bound for A_N, so there is an $n \geq N$ such that $x_n \leq x' + \frac{1}{2}\varepsilon < x' + \varepsilon$. Thus (ii)' holds.

Now suppose x' is a number satisfying (i)' and (ii)'. From (i)' it follows that $\inf A_n > x' - \varepsilon$ whenever $n \geq N$. Thus $\liminf x_n \geq x' - \varepsilon$, all ε, so $\liminf x_n \geq x'$. From (ii)' it follows that for any N and any ε, $\inf A_N < x' + \varepsilon$. Thus for any N, $\inf A_N \leq x'$, so $\liminf x_n \leq x'$. We have $\liminf x_n = x'$. □

Exercises

1. The sequence $(1/n)_{n=1}^{\infty}$ has limit 0. (Use the Archimedean axiom, §2.)
2. If $x_n > 0$ and $x_n \to 0$, then $x_n^{1/2} \to 0$.
3. If $a > 0$, then $a^{1/n} \to 1$ as $n \to \infty$. (Hint: if $a \geq 1$, let $a^{1/n} = 1 + x_n$. By the binomial expansion, or by induction, $a = (1 + x_n)^n \geq 1 + nx_n$. Thus $x_n < n^{-1}a \to 0$. If $a < 1$, then $a^{1/n} = (b^{1/n})^{-1}$ where $b = a^{-1} > 1$.)

4. $\lim n^{1/n} = 1$. (Hint: let $n^{1/n} = 1 + y_n$. For $n \geq 2$, $n = (1 + y_n)^n \geq 1 + ny_n + \frac{1}{2}n(n-1)y_n^2 > \frac{1}{2}n(n-1)y_n^2$, so $y_n^2 \leq 2(n-1)^{-1} \to 0$. Thus $y_n \to 0$.)

5. If $z \in \mathbb{C}$ and $|z| < 1$, then $z^n \to 0$ as $n \to \infty$.

6. Suppose $(x_n)_{n=1}^{\infty}$ is a bounded real sequence. Show that $x_n \to x$ if and only if $\lim \inf x_n = x = \lim \sup x_n$.

7. Prove the second part of Proposition 3.4.

8. Suppose $(x_n)_{n=1}^{\infty}$ and $(a_n)_{n=1}^{\infty}$ are two bounded real sequences such that $a_n \to a > 0$. Then

$$\lim \inf a_n x_n = a \cdot \lim \inf x_n, \qquad \lim \sup a_n x_n = a \cdot \lim \sup x_n.$$

§4. Series

Suppose $(z_n)_{n=1}^{\infty}$ is a sequence in $\mathbb{C}$. We associate to it a second sequence $(s_n)_{n=1}^{\infty}$, where

$$s_n = \sum_{m=1}^{n} z_m = z_1 + z_2 + \cdots + z_n.$$

If $(s_n)_{n=1}^{\infty}$ converges to s, it is reasonable to consider s as the infinite sum $\sum_{n=1}^{\infty} z_n$. Whether $(s_n)_{n=1}^{\infty}$ converges or not, the formal symbol $\sum_{n=1}^{\infty} z_n$ or $\sum z_n$ is called an *infinite series*, or simply a *series*. The number z_n is called the nth *term* of the series, s_n is called the nth *partial sum*. If $s_n \to s$ we say that the series $\sum z_n$ *converges* and that its *sum* is s. This is written

(4.1) $$s = \sum_{n=1}^{\infty} z_n.$$

(Of course if the sequence is indexed differently, e.g., $(z_n)_{n=0}^{\infty}$, we make the corresponding changes in defining s_n and in (4.1).) If the sequence $(s_n)_{n=1}^{\infty}$ does not converge, the series $\sum z_n$ is said to *diverge*.

In particular, suppose $(x_n)_{n=1}^{\infty}$ is a real sequence, and suppose each $x_n \geq 0$. Then the sequence $(s_n)_{n=1}^{\infty}$ of partial sums is clearly an increasing sequence. Either it is bounded, so (by Proposition 3.2) convergent, *or* for each $M > 0$ there is an N such that

$$s_n = \sum_{m=1}^{n} x_m > M \qquad \text{whenever } n \geq N.$$

In the first case we write

(4.2) $$\sum_{n=1}^{\infty} x_n < \infty$$

and in the second case we write

(4.3) $$\sum_{n=1}^{\infty} x_n = \infty.$$

Thus (4.2) $\Leftrightarrow$ $\sum x_n$ converges, (4.3) $\Leftrightarrow$ $\sum x_n$ diverges.

Examples

1. Consider the series $\sum_{n=1}^{\infty} n^{-1}$. We claim $\sum_{n=1}^{\infty} n^{-1} = \infty$. In fact (symbolically),

$$\sum n^{-1} = 1 + \tfrac{1}{2} + \tfrac{1}{3} + \tfrac{1}{4} + \tfrac{1}{5} + \tfrac{1}{6} + \tfrac{1}{7} + \tfrac{1}{8} + \cdots$$
$$\geq \tfrac{1}{2} + \tfrac{1}{2} + (\tfrac{1}{4} + \tfrac{1}{4}) + (\tfrac{1}{8} + \tfrac{1}{8} + \tfrac{1}{8} + \tfrac{1}{8}) + \cdots$$
$$= \tfrac{1}{2} + \tfrac{1}{2} + 2(\tfrac{1}{4}) + 4(\tfrac{1}{8}) + 8(\tfrac{1}{16}) + \cdots$$
$$= \tfrac{1}{2} + \tfrac{1}{2} + \tfrac{1}{2} + \cdots = \infty.$$

2. $\sum_{n=1}^{\infty} n^{-2} < \infty$. In fact (symbolically),

$$\sum n^{-2} = 1 + (\tfrac{1}{2})^2 + (\tfrac{1}{3})^2 + \cdots + (\tfrac{1}{7})^2 + \cdots$$
$$\leq 1 + (\tfrac{1}{2})^2 + (\tfrac{1}{2})^2 + (\tfrac{1}{4})^2 + (\tfrac{1}{4})^2 + (\tfrac{1}{4})^2 + (\tfrac{1}{4})^2 + \cdots$$
$$= 1 + 2(\tfrac{1}{2})^2 + 4(\tfrac{1}{4})^2 + 8(\tfrac{1}{8})^2 + \cdots$$
$$= 1 + \tfrac{1}{2} + \tfrac{1}{4} + \tfrac{1}{8} + \cdots = 2.$$

(We leave it to the reader to make the above rigorous by considering the respective partial sums.)

How does one tell whether a series converges? The question is whether the *sequence* $(s_n)_{n=1}^{\infty}$ of partial sums converges. Theorem 2.3 gives a necessary and sufficient condition for convergence of this sequence: that it be a Cauchy sequence. However this only refines our original question to: how does one tell whether a series has a sequence of partial sums which is a Cauchy sequence? The five propositions below give some answers.

Proposition 4.1. *If $\sum_{n=1}^{\infty} z_n$ converges, then $z_n \to 0$.*

Proof. If $\sum z_n$ converges, then the sequence $(s_n)_{n=1}^{\infty}$ of partial sums is a Cauchy sequence, so $s_n - s_{n-1} \to 0$. But $s_n - s_{n-1} = z_n$. $\square$

Note that the converse is false: $1/n \to 0$ but $\sum 1/n$ diverges.

Proposition 4.2. *If $|z| < 1$, then $\sum_{n=0}^{\infty} z^n$ converges; the sum is $(1 - z)^{-1}$. If $|z| \geq 1$, then $\sum_{n=0}^{\infty} z^n$ diverges.*

Proof. The nth partial sum is

$$s_n = 1 + z + z^2 + \cdots + z^{n-1}.$$

Then $s_n(1 - z) = 1 - z^n$, so $s_n = (1 - z^n)/(1 - z)$. If $|z| < 1$, then as $n \to \infty$, $z^n \to 0$ (Exercise 5 of §3). Therefore $s_n \to (1 - z)^{-1}$. If $|z| \geq 1$, then $|z^n| \geq 1$, and Proposition 4.1 shows divergence. $\square$

The series $\sum_{n=0}^{\infty} z^n$ is called a *geometric series*.

Proposition 4.3. (Comparison test). *Suppose $(z_n)_{n=1}^{\infty}$ is a sequence in $\mathbb{C}$ and $(a_n)_{n=1}^{\infty}$ a sequence in $\mathbb{R}$ with each $a_n \geq 0$. If there are constants M, N such that*

$$|z_n| \leq M a_n \quad \text{whenever } n \geq N,$$

and if $\sum a_n$ converges, then $\sum z_n$ converges.

Proof. Let $s_n = \sum_{m=1}^n z_m$, $b_n = \sum_{m=1}^n a_n$. If $n, m \geq N$ then

$$|s_n - s_m| = \left| \sum_{j=m+1}^n z_j \right| \leq \sum_{j=m+1}^n |z_j|$$

$$\leq M \sum_{j=m+1}^n a_n = M(b_n - b_m).$$

But $(b_n)_{n=1}^\infty$ is a Cauchy sequence, so this inequality implies that $(s_n)_{n=1}^\infty$ is also a Cauchy sequence. □

Proposition 4.4. (Ratio test). *Suppose $(z_n)_{n=1}^\infty$ is a sequence in $\mathbb{C}$ and suppose $z_n \neq 0$, all n.*

(a) *If*

$$\lim \sup |z_{n+1}/z_n| < 1,$$

then $\sum z_n$ converges.

(b) *If*

$$\lim \inf |z_{n+1}/z_n| > 1,$$

then $\sum z_n$ diverges.

Proof. (a) In this case, take r so that $\lim \sup |z_{n+1}/z_n| < r < 1$. By Proposition 3.4, there is an N so that $|z_{n+1}/z_n| \leq r$ whenever $n \geq N$. Thus if $n > N$,

$$|z_n| \leq r|z_{n-1}| \leq r \cdot r|z_{n-2}| \leq \cdots \leq r^{n-N}|z_N| = Mr^n,$$

where $M = r^{-N}|z_N|$. Propositions 4.2 and 4.3 imply convergence.

(b) In this case, Proposition 3.4 implies that for some N, $|z_{n+1}/z_n| \geq 1$ if $n \geq N$. Thus for $n > N$.

$$|z_n| \geq |z_{n-1}| \geq \cdots \geq |z_N| > 0.$$

We cannot have $z_n \to 0$, so Proposition 4.1 implies divergence. □

Corollary 4.5. *If $z_n \neq 0$ for $n = 1, 2, \ldots$ and if $\lim |z_{n+1}/z_n|$ exists, then the series $\sum z_n$ converges if the limit is < 1 and diverges if the limit is > 1.*

Note that for both the series $\sum 1/n$ and $\sum 1/n^2$, the limit in Corollary 4.5 equals 1. Thus either convergence or divergence is possible in this case.

Proposition 4.6. (Root test). *Suppose $(z_n)_{n=1}^\infty$ is a sequence in $\mathbb{C}$.*

(a) *If*

$$\lim \sup |z_n|^{1/n} < 1,$$

then $\sum z_n$ converges.

(b) *If*

$$\lim \sup |z_n|^{1/n} > 1,$$

then $\sum z_n$ diverges.

Proof. (a) In this case, take r so that $\limsup |z_n|^{1/n} < r < 1$. By Proposition 3.4, there is an N so that $|z_n|^{1/n} \le r$ whenever $n \ge N$. Thus if $n \ge N$, then $|z_n| \le r^n$. Propositions 4.2 and 4.3 imply convergence.

(b) In this case, Proposition 3.4 implies that $|z_n|^{1/n} \ge 1$ for infinitely many values of n. Thus Proposition 4.1 implies divergence. $\Box$

Note the tacit assumption in the statement and proof that $(|z_n|^{1/n})_{n=1}^{\infty}$ is a *bounded* sequence, so that the upper and lower limits exist. However, if this sequence is not bounded, then in particular $|z_n| \ge 1$ for infinitely many values of n, and Proposition 4.1 implies divergence.

Corollary 4.7. *If* $\lim |z_n|^{1/n}$ *exists, then the series* $\sum z_n$ *converges if the limit is* < 1 *and diverges if the limit is* > 1.

Note that for both the series $\sum 1/n$ and $\sum 1/n^2$, the limit in Corollary 4.7 equals 1 (see Exercise 4 of §3). Thus either convergence or divergence is possible in this case.

A particularly important class of series are the *power series*. If $(a_n)_{n=0}^{\infty}$ is a sequence in $\mathbb{C}$ and z_0 a fixed element of $\mathbb{C}$, then the series

$$(4.2) \qquad \sum_{n=0}^{\infty} a_n(z - z_0)^n$$

is the *power series around* z_0 with *coefficients* $(a_n)_{n=0}^{\infty}$. Here we use the convention that $w^0 = 1$ for all $w \in \mathbb{C}$, including $w = 0$. Thus (4.2) is defined, as a series, for each $z \in \mathbb{C}$. For $z = z_0$ it converges (with sum a_0), but for other values of z it may or may not converge.

Theorem 4.8. *Consider the power series* (4.2). *Define* R *by*

$$R = 0 \qquad \textit{if } (|a_n|^{1/n})_{n=1}^{\infty} \textit{ is not a bounded sequence},$$
$$R = (\limsup |a_n|^{1/n})^{-1} \qquad \textit{if } \limsup |a_n|^{1/n} > 0,$$
$$R = \infty \qquad \textit{if } \limsup |a_n|^{1/n} = 0.$$

Then the power series (4.2) *converges if* $|z - z_0| < R$, *and diverges if* $|z - z_0| > R$.

Proof. We have

$$(4.3) \qquad |a_n(z - z_0)^n|^{1/n} = |a_n|^{1/n}|z - z_0|.$$

Suppose $z \ne z_0$. If $(|a_n|^{1/n})_{n=1}^{\infty}$ is not a bounded sequence, then neither is (4.3), and we have divergence. Otherwise the conclusions follow from (4.3) and the root test, Proposition 4.6. $\Box$

The number R defined in the statement of Theorem 4.8 is called the *radius of convergence* of the power series (4.2). It is the radius of the largest circle in the complex plane inside which (4.2) converges.

Theorem 4.8 is quite satisfying from a theoretical point of view: the radius of convergence is shown to exist and is (in principle) determined in all cases. However, recognizing $\limsup |a_n|^{1/n}$ may be very difficult in practice. The following is often helpful.

Theorem 4.9. *Suppose* $a_n \neq 0$ *for* $n \geq N$, *and suppose* $\lim |a_{n+1}/a_n|$ *exists. Then the radius of convergence* R *of the power series* (4.2) *is given by*

$$R = (\lim |a_{n+1}/a_n|)^{-1} \quad \text{if } \lim |a_{n+1}/a_n| > 0.$$
$$R = \infty \quad \text{if } \lim |a_{n+1}/a_n| = 0.$$

Proof. Apply Corollary 4.5 to the series (4.2), noting that if $z \neq z_0$ then

$$|a_{n+1}(z - z_0)^{n+1}/a_n(z - z_0)^n| = |a_{n+1}/a_n| \cdot |z - z_0|. \qquad \Box$$

Exercises

1. If $\sum_{n=1}^{\infty} z_n$ converges with sum s and $\sum_{n=1}^{\infty} w_n$ converges with sum t, then $\sum_{n=1}^{\infty} (z_n + w_n)$ converges with sum $s + t$.

2. Suppose $\sum a_n$ and $\sum b_n$ each have all non-negative terms. If there are constants $M > 0$ and N such that $b_n \geq Ma_n$ whenever $n \geq N$, and if $\sum a_n = \infty$, then $\sum b_n = \infty$.

3. Show that $\sum_{n=1}^{\infty} (n + 1)/(2n^2 + 1)$ diverges and $\sum_{n=1}^{\infty} (n + 1)/(2n^3 + 1)$ converges. (Hint: use Proposition 4.3 and Exercise 2, and compare these to $\sum 1/n$, $\sum 1/n^2$.)

4. (2^k-Test). Suppose $a_1 \geq a_2 \geq \cdots \geq a_n \geq 0$, all n. Then $\sum_{n=1}^{\infty} a_n < \infty \Leftrightarrow \sum_{k=1}^{\infty} 2^k a_{2^k} < \infty$. (Hint: use the methods used to show divergence of $\sum 1/n$ and convergence of $\sum 1/n^2$.)

5. (Integral Test). Suppose $a_1 \geq a_2 \geq \cdots \geq a_n \geq 0$, all n. Suppose $f: [1, \infty) \to \mathbb{R}$ is a continuous function such that $f(n) = a_n$, all n, and $f(y) \leq f(x)$ if $y \geq x$. Then $\sum_{n=1}^{\infty} a_n < \infty \Leftrightarrow \int_1^{\infty} f(x)\, dx < \infty$.

6. Suppose $p > 0$. The series $\sum_{n=1}^{\infty} n^{-p}$ converges if $p > 1$ and diverges if $p \leq 1$. (Use Exercise 4 or Exercise 5.)

7. The series $\sum_{n=2}^{\infty} n^{-1}(\log n)^{-2}$ converges; the series $\sum_{n=2}^{\infty} n^{-1}(\log n)^{-1}$ diverges.

8. The series $\sum_{n=0}^{\infty} z^n/n!$ converges for any $z \in \mathbb{C}$. (Here $0! = 1$, $n! = n(n - 1)(n - 2) \cdot \ldots \cdot 1$.)

9. Determine the radius of convergence of

$$\sum_{n=0}^{\infty} 2^n z^n/n, \qquad \sum_{n=1}^{\infty} n^n z^n/n!, \qquad \sum_{n=0}^{\infty} n!\, z^n,$$

$$\sum_{n=0}^{\infty} n!\, z^n/(2n)!$$

10. (Alternating series). Suppose $|x_1| \geq |x_2| \geq \cdots \geq |x_n|$, all n, $x_n \geq 0$ if n odd, $x_n \leq 0$ if n even, and $x_n \to 0$. Then $\sum x_n$ converges. (Hint: the partial sums satisfy $s_2 \leq s_4 \leq s_6 \leq \cdots \leq s_5 \leq s_3 \leq s_1$.)

11. $\sum_{n=1}^{\infty} (-1)^n/n$ converges.

§5. Metric spaces

A *metric* on a set S is a function d from the product set $S \times S$ to $\mathbb{R}$, with the properties

D1. $d(x, x) = 0, d(x, y) > 0$, if $x, y \in S, x \neq y$.
D2. $d(x, y) = d(y, x)$, all $x, y \in S$.
D3. $d(x, z) \leq d(x, y) + d(y, z)$, all $x, y, z \in S$.

We shall refer to $d(x, y)$ as the *distance* from x to y. A *metric space* is a set S together with a given metric d. The inequality D3 is called the *triangle inequality*. The elements of S are often called *points*.

As an example, take $S = \mathbb{R}^2 = \mathbb{R} \times \mathbb{R}$, with

(5.1) $\qquad d((x, y), (x', y')) = [(x - x')^2 + (y - y')^2]^{1/2}$.

If we coordinatize the Euclidean plane in the usual way, and if $(x, y), (x', y')$ are the coordinates of points P and P' respectively, then (5.1) gives the length of the line segment PP' (Pythagorean theorem). In this example, D3 is the analytic expression of the fact that the length of one side of a triangle is at most the sum of the lengths of the other two sides. The same example in different guise is obtained by letting $S = \mathbb{C}$ and taking

(5.2) $\qquad d(z, w) = |z - w|$

as the metric. Then D3 is a consequence of Exercise 5 in §2.

Some other possible metrics on $\mathbb{R}^2$ are:

$d_1((x, y), (x', y')) = |x - x'| + |y - y'|$,
$d_2((x, y), (x', y')) = \max \{|x - x'|, |y - y'|\}$,
$d_3((x, y), (x', y')) = 0 \qquad$ if $(x, y) = (x', y')$, and 1 otherwise.

Verification that the functions d_1, d_2, and d_3 satisfy the conditions D1, D2, D3 is left as an exercise. Note that d_3 works for *any* set S: if $x, y \in S$ we set $d(x, y) = 1$ if $x \neq y$ and 0 if $x = y$.

A still simpler example of a metric space is $\mathbb{R}$, with distance function d given by

(5.3) $\qquad d(x, y) = |x - y|$.

Again this coincides with the usual notion of the distance between two points on the (coordinatized) line.

Another important example is $\mathbb{R}^n$, the space of ordered n-tuples $x = (x_1, x_2, \ldots, x_n)$ of elements of $\mathbb{R}$. There are various possible metrics on $\mathbb{R}^n$ like the metrics d_1, d_2, d_3 defined above for $\mathbb{R}^n$, but we shall consider here only the generalization of the Euclidean distance in $\mathbb{R}^2$ and $\mathbb{R}^3$. If $x = (x_1, x_2, \ldots, x_n)$ and $y = (y_1, y_2, \ldots, y_n)$ we set

(5.4) $\qquad d(x, y) = [(x_1 - y_1)^2 + (x_2 - y_2)^2 + \cdots + (x_n - y_n)^2]^{1/2}$.

When $n = 1$ we obtain $\mathbb{R}$ with the metric (5.3); when $n = 2$ we obtain $\mathbb{R}^2$ with the metric (5.1), in somewhat different notation. It is easy to verify that

d given by (5.4) satisfies D1 and D2, but condition D3 is not so easy to verify. For now we shall simply *assert* that d satisfies D3; a proof will be given in a more general setting in Chapter 4.

Often when the metric d is understood, one refers to a set S alone as a metric space. For example, when we refer to $\mathbb{R}$, $\mathbb{C}$, or $\mathbb{R}^n$ as a metric space with no indication what metric is taken, we mean the metric to be given by (5.3), (5.2), or (5.4) respectively.

Suppose (S, d) is a metric space and T is a subset of S. We can consider T as a metric space by taking the distance function on $T \times T$ to be the restriction of d to $T \times T$.

The concept of metric space has been introduced to provide a uniform treatment of such notions as distance, convergence, and limit which occur in many contexts in analysis. Later we shall encounter metric spaces much more exotic than $\mathbb{R}^n$ and $\mathbb{C}$.

Suppose (S, d) is a metric space, x is a point of S, and r is a positive real number. The *ball of radius r about* x is defined to be the subset of S consisting of all points in S whose distance from x is less than r:

$$B_r(x) = \{y \mid y \in S, d(x, y) < r\}.$$

Clearly $x \in B_r(x)$. If $0 < r < s$, then $B_r(x) \subset B_s(x)$.

Examples

When $S = \mathbb{R}$ (metric understood), $B_r(x)$ is the open interval $(x - r, x + r)$. When $S = \mathbb{R}^2$ or $\mathbb{C}$, $B_r(z)$ is the open disc of radius r centered at z. Here we take the adjective "open" as understood; we shall see that the interval and the disc in question are also open in the sense defined below.

A subset $A \subset S$ is said to be a *neighborhood* of the point $x \in S$ if A contains $B_r(x)$ for *some* $r > 0$. Roughly speaking, this says that A contains all points sufficiently close to x. In particular, if A is a neighborhood of x it contains x itself.

A subset $A \subset S$ is said to be *open* if it is a neighborhood of each of its points. Note that the empty set is an open subset of S: since it has no points (elements), it is a neighborhood of each one it has.

Example

Consider the interval $A = (0, 1] \subset \mathbb{R}$. This is a neighborhood of each of its points except $x = 1$. In fact, if $0 < x < 1$, let $r = \min\{x, 1 - x\}$. Then $A \supset B_r(x) = (x - r, x + r)$. However, for *any* $r > 0$, $B_r(1)$ contains $1 + \frac{1}{2}r$, which is not in A.

We collect some useful facts about open sets in the following proposition.

Proposition 5.1. *Suppose (S, d) is a metric space.*

(a) *For any $x \in S$ and any $r > 0$, $B_r(x)$ is open.*

(b) *If $A_1, A_2, \ldots, A_n$ are open subsets of S, then $\bigcap_{m=1}^{n} A_m$ are also open.*

(c) *If $(A_\beta)_{\beta \in B}$ is any collection of open subsets of S, then $\bigcup_{\beta \in B} A_\beta$ is also open.*

Proof. (a) Suppose $y \in B_r(x)$. We want to show that for some $s > 0$, $B_s(y) \subset B_r(x)$. The triangle inequality makes this easy, for we can choose $s = r - d(y, x)$. (Since $y \in B_r(x)$, s is positive.) If $z \in B_s(y)$, then

$$d(z, x) \leq d(z, y) + d(y, x) < s + d(y, x) = r.$$

Thus $z \in B_r(x)$.

(b) Suppose $x \in \bigcap_{m=1}^{n} A_m$. Since each A_m is open, there is $r(m) > 0$ so that $B_{r(m)}(x) \subset A_m$. Let $r = \min \{r(1), r(2), \ldots, r(n)\}$. Then $r > 0$ and $B_r(x) \subset B_{r(m)}(x) \subset A_m$, so $B_r(x) \subset \bigcap_{m=1}^{n} A_m$. (Why is it necessary here to assume that $A_1, A_2, \ldots$ is a *finite* collection of sets?)

(c) Suppose $x \in A = \bigcup_{\beta \in B} A_\beta$. Then for some particular β, $x \in A_\beta$. Since A_β is open, there is an $r > 0$ so that $B_r(x) \subset A_\beta \subset A$. Thus A is open. $\Box$

Again suppose (S, d) is a metric space and suppose $A \subset S$. A point $x \in S$ is said to be a *limit point of A* if for every $r > 0$ there is a point of A with distance from x less than r:

$$B_r(x) \cap A \neq \varnothing \qquad \text{if } r > 0.$$

In particular, if $x \in A$ then x is a limit point of A. The set A is said to be *closed* if it contains each of its limit points. Note that the empty set is closed, since it has no limit points.

Example

The interval $(0, 1] \subset \mathbb{R}$ has as its set of limit points the closed interval $[0, 1]$. In fact if $0 < x \leq 1$, then x is certainly a limit point. If $x = 0$ and $r > 0$, then $B_r(0) \cap (0, 1] = (-r, r) \cap (0, 1] \neq \varnothing$. If $x < 0$ and $r = |x|$, then $B_r(x) \cap (0, 1] = \varnothing$, while if $x > 1$ and $r = x - 1$, then $B_r(x) \cap (0, 1] = \varnothing$. Thus the interval $(0, 1]$ is *neither open nor closed*. The exact relationship between open sets and closed sets is given in Proposition 5.3 below.

The following is the analogue for closed sets of Proposition 5.1.

Proposition 5.2. *Suppose (S, d) is a metric space.*

(a) *For any $x \in S$ and any $r > 0$, the closed ball $C = \{y \mid y \in S, d(x, y) \leq r\}$ is a closed set.*

(b) *If $A_1, A_2, \ldots, A_n$ are closed subsets of S, then $\bigcup_{m=1}^{n} A_m$ is closed.*

(c) *If $(A_\beta)_{\beta \in B}$ is any collection of closed subsets of S, then $\bigcap_{\beta \in B} A_\beta$ is closed.*

Proof. (a) Suppose z is a limit point of the set C. Given $\varepsilon > 0$, there is a point $y \in B_\varepsilon(z) \cap C$. Then

$$d(z, x) \leq d(z, y) + d(y, x) < \varepsilon + r.$$

Since this is true for every $\varepsilon > 0$, we must have $d(z, x) \leq r$. Thus $z \in C$.

(b) Suppose $x \notin A = \bigcup_{m=1}^{n} A_m$. For each m, x is not a limit point of A_m, so there is $r(m) > 0$ such that $B_{r(m)}(x) \cap A_m = \varnothing$. Let

$$r = \min \{r(1), r(2), \ldots, r(n)\}.$$

Then $B_r(x) \cap A_m = \varnothing$, all m, so $B_r(x) \cap A = \varnothing$. Thus x is not a limit point of A.

(c) Suppose x is a limit point of $A = \bigcap_{\beta \in B} A_\beta$. For any $r > 0$, $B_r(x) \cap A \neq \varnothing$. But $A \subset A_\beta$, so $B_r(x) \cap A_\beta \neq \varnothing$. Thus x is a limit point of A_β, so it is in A_β. This is true for each β, so $x \in A$. □

Proposition 5.3. *Suppose (S, d) is a metric space. A subset $A \subset S$ is open if and only if its complement is closed.*

Proof. Let B be the complement of A. Suppose B is closed, and suppose $x \in A$. Then x is not a limit point of B, so for some $r > 0$ we have $B_r(x) \cap B = \varnothing$. Thus $B_r(x) \subset A$, and A is a neighborhood of x.

Conversely, suppose A is open and suppose $x \notin B$. Then $x \in A$, so for some $r > 0$ we have $B_r(x) \subset A$. Then $B_r(x) \cap B = \varnothing$, and x is not a limit point of B. It follows that every limit point of B is in B. □

The set of limit points of a subset $A \subset S$ is called the *closure* of A; we shall denote it by A^-. We have $A \subset A^-$ and A is closed if and only if $A = A^-$. In the example above, we saw that the closure of $(0, 1] \subset \mathbb{R}$ is $[0, 1]$.

Suppose A, B are subsets of S and $A \subset B$. We say that A is *dense in* B if $B \subset A^-$. In particular, A is dense in S if $A^- = S$. As an example, $\mathbb{Q}$ (the rationals) is dense in $\mathbb{R}$. In fact, suppose $x \in \mathbb{R}$ and $r > 0$. Choose a positive integer n so large that $1/n < r$. There is a unique integer m so that $m/n < x < (m + 1)/n$. Then $d(x, m/n) = x - m/n < (m + 1)/n - m/n = 1/n < r$, so $m/n \in B_r(x)$. Thus $x \in \mathbb{Q}^-$.

A sequence $(x_n)_{n=1}^{\infty}$ in S is said to *converge to* $x \in S$ if for each $\varepsilon > 0$ there is an N so that $d(x_n, x) < \varepsilon$ if $n \geq N$. The point x is called the *limit* of the sequence, and we write

$$\lim_{n \to \infty} x_n = x \quad \text{or} \quad x_n \to x.$$

When $S = \mathbb{R}$ or $\mathbb{C}$ (with the usual metric), this coincides with the definition in §3. Again the limit, if any, is unique.

A sequence $(x_n)_{n=1}^{\infty}$ in S is said to be a *Cauchy sequence* if for each $\varepsilon > 0$ there is an N so that $d(x_n, x_m) < \varepsilon$ if $n, m \geq N$. Again when $S = \mathbb{R}$ or $\mathbb{C}$, this coincides with the definition in §3.

The metric space (S, d) is said to be *complete* if every Cauchy sequence in S converges to a point of S. As an example, Theorem 3.3 says precisely that $\mathbb{R}$ and $\mathbb{C}$ are complete metric spaces with respect to the usual metrics.

Many processes in analysis produce sequences of numbers, functions, etc., in various metric spaces. It is important to know when such sequences converge. Knowing that the metric space in question is complete is a powerful tool, since the condition that the sequence be a Cauchy sequence is then a necessary and sufficient condition for convergence. We have already seen this in our discussion of series, for example.

Note that $\mathbb{R}^n$ is complete. To see this note that in $\mathbb{R}^n$,

$$\max\{|x_j - y_j|, j = 1, \ldots, n\} \leq d(x, y) \leq n \cdot \max\{|x_j - y_j|, j = 1, \ldots, n\}.$$

It follows that a sequence of points in $\mathbb{R}^n$ converges if and only if each of the n corresponding sequences of coordinates converges in $\mathbb{R}$. Similarly, a sequence of points in $\mathbb{R}^n$ is a Cauchy sequence if and only if each of the n corresponding sequences of coordinates is a Cauchy sequence in $\mathbb{R}$. Thus completeness of $\mathbb{R}^n$ follows from completeness of $\mathbb{R}$. (This is simply a generalization of the argument showing $\mathbb{C}$ is complete.)

Exercises

1. If (S, d) is a metric space, $x \in S$, and $r \geq 0$, then

$$\{y \mid y \in S, d(y, x) > r\}$$

is an open subset of S.

2. The point x is a limit point of a set $A \subset S$ if and only if there is a sequence $(x_n)_{n=1}^{\infty}$ in A such that $x_n \rightarrow x$.

3. If a sequence $(x_n)_{n=1}^{\infty}$ in a metric space converges to $x \in S$ and also converges to $y \in S$, then $x = y$.

4. If a sequence converges, then it is a Cauchy sequence.

5. If (S, d) is a complete metric space and $A \subset S$ is closed, then (A, d) is complete. Conversely, if $B \subset S$ and (B, d) is complete, then B is a closed subset of S.

6. The interval $(0, 1)$ is open as a subset of $\mathbb{R}$, but *not* as a subset of $\mathbb{C}$.

7. Let $S = \mathbb{Q}$ (the rational numbers) and let $d(x, y) = |x - y|$, $x, y \in \mathbb{Q}$. Show that (S, d) is not complete.

8. The set of all elements $x = (x_1, x_2, \ldots, x_n)$ in $\mathbb{R}^n$ such that each x_j is rational is a dense subset of $\mathbb{R}^n$.

9. Verify that $\mathbb{R}^n$ is complete.

§6. Compact Sets

Suppose that (S, d) is a metric space, and suppose A is a subset of S. The subset A is said to be *compact* if it has the following property: suppose that for each $x \in A$ there is given a neighborhood of x, denoted $N(x)$; then there are finitely many points $x_1, x_2, \ldots, x_n$ in A such that A is contained in the union of $N(x_1), N(x_2), \ldots, N(x_n)$. (Note that we are saying that this is true for *any* choice of neighborhoods of points of A, though the selection of points $x_1, x_2, \ldots$ may depend on the selection of neighborhoods.) It is obvious that any *finite* subset A is compact.

Examples

1. The infinite interval $(0, \infty) \subset \mathbb{R}$ is not compact. For example, let $N(x) = (x - 1, x + 1)$, $x \in (0, \infty)$. Clearly no finite collection of these intervals of finite length can cover all of $(0, \infty)$.

2. Even the finite interval $(0, 1] \subset \mathbb{R}$ is not compact. To see this, let $N(x) = (\frac{1}{2}x, 2)$, $x \in (0, 1]$. For any $x_1, x_2, \ldots, x_n \in (0, 1]$, the union of the intervals $N(x_i)$ will not contain y if $y \leq \frac{1}{2} \min \{x_1, x_2, \ldots, x_n\}$.

3. The set $A = \{0\} \cup \{1, \frac{1}{2}, \frac{1}{3}, \frac{1}{4}, \ldots\} \subset \mathbb{R}$ is compact. In fact, suppose for each $x \in A$ we are given a neighborhood $N(x)$. In particular, the neighborhood $N(0)$ of 0 contains an interval $(-\varepsilon, \varepsilon)$. Let M be a positive integer larger than $1/\varepsilon$. Then $1/n \in N(0)$ for $n \geq M$, and it follows that $A \subset N(0) \cup N(1) \cup N(\frac{1}{2}) \cup \cdots \cup N(1/M)$.

The first two examples illustrate general requirements which compact sets must satisfy. A subset A of S, when (S, d) is a metric space, is said to be *bounded* if there is a ball $B_r(x)$ containing A.

Proposition 6.1. *Suppose (S, d) is a metric space, $S \neq \varnothing$, and suppose $A \subset S$ is compact. Then A is closed and bounded.*

Proof. Suppose $y \notin A$. We want to show that y is not a limit point of A. For any $x \in A$, let $N(x)$ be the ball of radius $\frac{1}{2}d(x, y)$ around x. By assumption, there are $x_1, x_2, \ldots, x_n \in A$ such that $A \subset \bigcup_{m=1}^{n} N(x_m)$. Let r be the minimum of the numbers $\frac{1}{2}d(x_1, y), \ldots, \frac{1}{2}d(x_n, y)$. If $x \in A$, then for some m, $d(x, x_m) < \frac{1}{2}d(x_m, y)$. But then

$$d(x_m, y) \leq d(x_m, x) + d(x, y)$$
$$< \frac{1}{2}d(x_m, y) + d(x, y).$$

so $d(x, y) > \frac{1}{2}d(x_m, y) \geq r$. Thus $B_r(y) \cap A = \varnothing$, and y is not a limit point of A.

Next, we want to show that A is bounded. For each $x \in A$, let $N(x)$ be the ball of radius 1 around x. Again, by assumption there are $x_1, x_2, \ldots, x_n \in A$ such that $A \subset \bigcup_{m=1}^{n} N(x_m)$. Let

$$r = 1 + \max \{d(x_1, x_2), d(x_1, x_3), \ldots, d(x_1, x_n)\}.$$

If $y \in A$ then for some m, $d(y, x_m) < 1$. Therefore $d(y, x_1) \leq d(y, x_m) + d(x_m, x_1) < 1 + d(x_m, x_1) \leq r$, and we have $A \subset B_r(x_1)$. $\square$

The converse of Proposition 6.1, that a closed, bounded subset of a metric space is compact, is not true in general. It is a subtle but extremely important fact that it *is* true in $\mathbb{R}^n$, however.

Theorem 6.2. (Heine-Borel Theorem). *A subset of $\mathbb{R}^n$ or of $\mathbb{C}$ is compact if and only if it is closed and bounded.*

Proof. We have seen that in any metric space, if A is compact it is necessarily closed and bounded. Conversely, suppose $A \subset \mathbb{R}^n$ is closed and bounded. Let us assume at first that $n = 1$. Since A is bounded, it is contained in some closed interval $[a, b]$. Suppose for each $x \in A$, we are given a neighborhood $N(x)$ of x. We shall say that a closed subinterval of $[a, b]$ is *nice* if there are points $x_1, x_2, \ldots, x_m \in A$ such that $\bigcup_{j=1}^{m} N(x_j)$ contains the intersection of the subinterval with A; we are trying to show that $[a, b]$ itself is nice. Suppose it is not. Consider the two subintervals $[a, c]$ and $[c, b]$, where $c = \frac{1}{2}(a + b)$ is the midpoint of $[a, b]$. If both of these were nice, it would follow that $[a, b]$ itself is nice. Therefore we must have one of them not nice; denote its endpoints by a_1, b_1, and let $c_1 = \frac{1}{2}(a_1 + b_1)$. Again, one of the intervals $[a_1, c_1]$ and $[c_1, b_1]$ must not be nice; denote it by $[a_2, b_2]$. Continuing in this way we get a sequence of intervals $[a_m, b_m]$, $m = 0, 1, 2, \ldots$ such that $[a_0, b_0] = [a, b]$, each $[a_m, b_m]$ is the left or right half of the interval $[a_{m-1}, b_{m-1}]$, and each interval $[a_m, b_m]$ is not nice. It follows that $a_0 \le a_1 \le \cdots \le a_m \le b_m \le \cdots \le b_1 \le b_0$ and $b_m - a_m = 2^{-m}(b_0 - a_0) \to 0$. Therefore there is a point x such that $a_m \to x$ and $b_m \to x$. Moreover, $a_m \le x \le b_m$, for all m. We claim that $x \in A$; it is here that we use the assumption that A is closed. Since $[a_m, b_m]$ is not nice, it must contain points of A: otherwise $A \cap [a_m, b_m] = \varnothing$ would be contained in *any* $\bigcup_{j=1}^{r} N(x_j)$. Let

$$x_m \in [a_m, b_m] \cap A.$$

Clearly $x_m \to x$, since $a_m \to x$ and $b_m \to x$. Since A is closed, we get $x \in A$. Now consider the neighborhood $N(x)$. This contains an interval $(x - \varepsilon, x + \varepsilon)$. If we choose m so large that $b_m - a_m < \varepsilon$, then since $a_m \le x \le b_m$ this implies $[a_m, b_m] \subset N(x)$. But this means that $[a_m, b_m]$ is nice. This contradiction proves the theorem for the case $n = 1$.

The same method of proof works in $\mathbb{R}^n$, where instead of intervals we use squares, cubes, or their higher dimensional analogues. For example, when $n = 2$ we choose M so large that A is contained in the square with corners $(\pm M, \pm M)$. If this square were not nice, the same would be true of one of the four equal squares into which it can be divided, and so on. Continuing we get a sequences of squares $S_0 \supset S_1 \supset S_2 \supset \cdots$, each of side $\frac{1}{2}$ the length of the preceding, each intersecting A, and each not nice. The intersection $\bigcap_{m=0}^{\infty} S_m$ contains a single point x, and x is in A. Then $N(x)$ contains S_m for large m, a contradiction. Since as metric space $\mathbb{C} = \mathbb{R}^2$, this also proves the result for $\mathbb{C}$. ☐

Suppose $(x_n)_{n=1}^{\infty}$ is a sequence in a set S. A *subsequence* of this sequence is a sequence of the form $(y_k)_{k=1}^{\infty}$, where for each k there is a positive integer n_k so that

$$n_1 < n_2 < \cdots < n_k < n_{k+1} < \cdots,$$
$$y_k = x_{n_k}.$$

Thus, $(y_k)_{k=1}^{\infty}$ is just a selection of some (possibly all) of the x_n's, taken in order. As an example, if $(x_n)_{n=1}^{\infty} \subset \mathbb{R}$ has $x_n = (-1)^n/n$, and if we take $n_k = 2k$, then $(x_n)_{n=1}^{\infty} = (-1, \frac{1}{2}, -\frac{1}{3}, \frac{1}{4}, -\frac{1}{5}, \ldots)$ and $(y_k)_{k=1}^{\infty} = (\frac{1}{2}, \frac{1}{4}, \frac{1}{6}, \ldots)$. As a second example, let $(x_n)_{n=1}^{\infty}$ be an enumeration of the rationals. Then for any real number x, there is a subsequence of $(x_n)_{n=1}^{\infty}$ which converges to x.

Suppose (S, d) is a metric space. A set $A \subset S$ is said to be *sequentially compact* if, given any sequence $(x_n)_{n=1}^{\infty} \subset A$, some subsequence converges to a point of A.

Examples

1. Any finite set is sequentially compact. (Prove this.)
2. The interval $(0, \infty) \subset \mathbb{R}$ is not sequentially compact; in fact let $x_n = n$. No subsequence of $(x_n)_{n=1}^{\infty}$ converges.
3. The bounded interval $(0, 1] \subset \mathbb{R}$ is not sequentially compact; in fact let $x_n = 1/n$. Any subsequence of $(x_n)_{n=1}^{\infty}$ converges to 0, which is not in $(0, 1]$.

Proposition 6.3. *Suppose (S, d) is a metric space, $S \neq \varnothing$, and suppose $A \subset S$ is sequentially compact. Then A is closed and bounded.*

Proof. Suppose x is a limit point of A. Choose $x_n \in B_{1/n}(x) \cap A$, $n = 1, 2, 3, \ldots$. Any subsequence of $(x_n)_{n=1}^{\infty}$ converges to x, since $x_n \to x$. It follows (since by assumption some subsequence converges to a point of A) that $x \in A$. Thus A is closed.

Suppose A were not bounded. Take $x \in S$ and choose $x_1 \in A$ such that $x_1 \notin B_1(x)$. Let $r_1 = d(x, x_1) + 1$. By the triangle inequality, $B_1(x_1) \subset B_{r_1}(x)$. Since A is not bounded, there is $x_2 \in A$ such that $x_2 \notin B_{r_1}(x)$. Thus also $d(x_1, x_2) \geq 1$. Let $r_2 = \max \{d(x, x_1), d(x, x_2)\} + 1$ and choose $x_3 \in A$ such that $x_3 \notin B_{r_2}(x)$. Then $d(x_1, x_3) \geq 1$ and $d(x_2, x_3) \geq 1$. Continuing in this way we can find a sequence $(x_n)_{n=1}^{\infty} \subset A$ such that $d(x_m, x_n) \geq 1$ if $m \neq n$. Then no subsequence of this sequence can converge, and A is not sequentially compact. $\square$

Theorem 6.4. (Bolzano-Weierstrass Theorem). *A subset A of $\mathbb{R}^n$ or of $\mathbb{C}$ is sequentially compact if and only if it is closed and bounded.*

Proof. We have shown that A sequentially compact implies A closed and bounded. Suppose A is closed and bounded, and suppose first that $n = 1$. Take an interval $[a, b]$ containing A. Let $c = \frac{1}{2}(a + b)$. One (or both) of the subintervals $[a, c]$ and $[c, b]$ must contain x_n for infinitely many integers n; denote such a subinterval by $[a_1, b_1]$, and consider $[a_1, c_1]$, $[c_1, b_1]$ where $c_1 = \frac{1}{2}(a_1 + b_1)$. Proceeding in this way we can find intervals $[a_m, b_m]$ with the properties $[a_0, b_0] = [a, b]$, $[a_m, b_m] \subset [a_{m-1}, b_{m-1}]$, $b_m - a_m = 2^{-m}(b_0 - a_0)$, and $[a_m, b_m]$ contains x_n for infinitely many values of n. Then there is a point x such that $a_m \to x$, $b_m \to x$. We choose integers $n_1, n_2, \ldots$ so that $x_{n_1} \in [a_1, b_1]$, $n_2 > n_1$ and $x_{n_2} \in [a_2, b_2]$, $n_3 > n_2$ and $x_{n_3} \in [a_3, b_3]$, etc. Then this subsequence converges to x. Since A is closed, $x \in A$.

The generalization of this proof to higher dimensions now follows as in the proof of Theorem 6.2. $\square$

Both the terminology and the facts proved suggest a close relationship between compactness and sequential compactness. This relationship is made precise in the exercises below.

Exercises

1. Suppose $(x_n)_{n=1}^{\infty}$ is a sequence in a metric space (S, d) which converges to $x \in S$. Let $A = \{x\} \cup \{x_n\}_{n=1}^{\infty}$. Then A is compact and sequentially compact.

2. Let $\mathbb{Q}$, the rationals, have the usual metric. Let $A = \{x \mid x \in \mathbb{Q}, x^2 < 2\}$. Then A is bounded, and is closed as a subset of $\mathbb{Q}$, but is not compact.

3. Suppose A is a compact subset of a metric space (S, d). Then A is sequentially compact. (Hint: otherwise there is a sequence $(x_n)_{n=1}^{\infty}$ in A with no subsequence converging to a point of A. It follows that for each $x \in A$. there is an $r(x) > 0$ such that the ball $N(x) = B_{r(x)}(x)$ contains x_n for only finitely many values of n. Since A is compact, this would imply that $\{1, 2, 3, \ldots\}$ is finite, a contradiction.)

4. A metric space is said to be *separable* if there is a dense subset which is countable. If (S, d) is separable and $A \subset S$ is sequentially compact, then A is compact. (Hint: suppose for each $x \in A$ we are given a neighborhood $N(x)$. Let $\{x_1, x_2, x_3, \ldots\}$ be a dense subset of S. For each $x \in A$ we can choose an integer m and a rational r_m such that $x \in B_{r_m}(x_m) \subset N(x)$. The collection of balls $B_{r_m}(x_m)$ so obtained is (finite or) countable; enumerate them as C_1, $C_2, \ldots$. Since each C_j is contained in some $N(x)$, it is sufficient to show that for some n, $\bigcup_{j=1}^{n} C_j \supset A$. If this were not the case, we could take $y_n \in A$, $y_n \notin \bigcup_{j=1}^{n} C_j$, $n = 1, 2, \ldots$. Applying the assumption of sequential compactness to this sequence and noting how the C_j were obtained, we get a contradiction.)

§7. Vector spaces

A vector space over $\mathbb{R}$ is a set $\mathbf{X}$ in which there are an operation of addition and an operation of multiplication by real numbers which satisfy certain conditions. These abstract from the well-known operations with directed line segments in Euclidean 3-space.

Specifically, we assume that there is a function from $\mathbf{X} \times \mathbf{X}$ to $\mathbf{X}$, called *addition*, which assigns to the ordered pair $(\mathbf{x}, \mathbf{y}) \in \mathbf{X} \times \mathbf{X}$ an element of $\mathbf{X}$ denoted $\mathbf{x} + \mathbf{y}$. We assume

V1. $(\mathbf{x} + \mathbf{y}) + \mathbf{z} = \mathbf{x} + (\mathbf{y} + \mathbf{z})$, all $\mathbf{x}, \mathbf{y}, \mathbf{z} \in \mathbf{X}$.
V2. $\mathbf{x} + \mathbf{y} = \mathbf{y} + \mathbf{x}$, all $\mathbf{x}, \mathbf{y} \in \mathbf{X}$.
V3. There exists $\mathbf{0} \in \mathbf{X}$ such that $\mathbf{x} + \mathbf{0} = \mathbf{x}$, all $\mathbf{X}$.
V4. For all $\mathbf{x} \in \mathbf{X}$, there exists $-\mathbf{x} \in \mathbf{X}$ such that $\mathbf{x} + (-\mathbf{x}) = \mathbf{0}$.

We assume also that there is a function from $\mathbb{R} \times \mathbf{X}$ to $\mathbf{X}$, called *scalar multiplication*, assigning to the ordered pair $(a, \mathbf{x}) \in \mathbb{R} \times \mathbf{X}$ an element of $\mathbf{X}$ denoted $a\mathbf{x}$. We assume

V5. $(ab)\mathbf{x} = a(b\mathbf{x})$, all $a, b \in \mathbb{R}$, $\mathbf{x} \in \mathbf{X}$.
V6. $a(\mathbf{x} + \mathbf{y}) = a\mathbf{x} + a\mathbf{y}$, all $a \in \mathbb{R}$, $\mathbf{x}, \mathbf{y} \in \mathbf{X}$.
V7. $(a + b)\mathbf{x} = a\mathbf{x} + b\mathbf{x}$, all $a, b \in \mathbb{R}$, $\mathbf{x} \in \mathbf{X}$.
V8. $1\mathbf{x} = \mathbf{x}$, all $\mathbf{x} \in \mathbf{X}$.

Summarizing: a *vector space over* $\mathbb{R}$, or a *real vector space*, is a set $\mathbf{X}$ with addition satisfying V1–V4 and scalar multiplication satisfying V5–V8. The elements of $\mathbf{X}$ are called *vectors* and the elements of $\mathbb{R}$, in this context, are often called *scalars*.

Similarly, a *vector space over* $\mathbb{C}$, or a *complex vector space*, is a set $\mathbf{X}$ together with addition satisfying V1–V4 and scalar multiplication defined from $\mathbb{C} \times \mathbf{X}$ to $\mathbf{X}$ and satisfying V5–V8. Here the scalars are, of course, *complex* numbers.

Examples

1. $\mathbb{R}$ is a vector space over $\mathbb{R}$, with addition as usual and the usual multiplication as scalar multiplication.

2. The set with one element $\mathbf{0}$ is a vector space over $\mathbb{R}$ or $\mathbb{C}$ with $\mathbf{0} + \mathbf{0} = \mathbf{0}$, $a\mathbf{0} = \mathbf{0}$, all a.

3. $\mathbb{R}^n$ is a vector space over $\mathbb{R}$ if we take addition and scalar multiplication as

$$(x_1, x_2, \ldots, x_n) + (y_1, y_2, \ldots, y_n) = (x_1 + y_1, x_2 + y_2, \ldots, x_n + y_n),$$
$$a(x_1, x_2, \ldots, x_n) = (ax_1, ax_2, \ldots, ax_n).$$

4. $\mathbb{C}$ is a vector space over $\mathbb{R}$ *or* $\mathbb{C}$ with the usual addition and scalar multiplication.

5. Let S be any set, and let $F(S; \mathbb{R})$ be the set whose elements s are the functions from S to $\mathbb{R}$. Define addition and scalar multiplication in $F(S; \mathbb{R})$ by

$$(f + g)(s) = f(s) + g(s), \qquad s \in S,$$
$$(af)(s) = af(s), \qquad s \in S, a \in \mathbb{R}.$$

Then $F(S; \mathbb{R})$ is a vector space over $\mathbb{R}$.

6. The set $F(S; \mathbb{C})$ of functions from S to $\mathbb{C}$ can be made a complex vector space by defining addition and scalar multiplication as in 5.

7. Let $\mathbf{X}$ be the set of all functions $f: \mathbb{R} \to \mathbb{R}$ which are polynomials, i.e., for some $a_0, a_1, \ldots, a_n \in \mathbb{R}$,

$$f(x) = a_0 + a_1 + a_2 x^2 + \cdots + a_n x^n, \qquad \text{all } x \in \mathbb{R}.$$

With addition and scalar multiplication defined as in 5, this is a real vector space.

8. The set of polynomials with complex coefficients can be considered as a complex vector space.

Let us note two elementary facts valid in every vector space: the element $\mathbf{0}$ of assumption V3 is unique, and for any $\mathbf{x} \in \mathbf{X}$, $0\mathbf{x} = \mathbf{0}$. First, suppose $\mathbf{0}' \in \mathbf{X}$ has the property that $\mathbf{x} + \mathbf{0}' = \mathbf{x}$ for each $\mathbf{x} \in \mathbf{X}$. Then in particular $\mathbf{0}' = \mathbf{0}' + \mathbf{0} = \mathbf{0} + \mathbf{0}' = \mathbf{0}$ (using V2 and V3). Next, if $\mathbf{x} \in \mathbf{X}$, then

$$
\begin{aligned}
0\mathbf{x} &= 0\mathbf{x} + \mathbf{0} = 0\mathbf{x} + [0\mathbf{x} + (-0\mathbf{x})] \\
&= [0\mathbf{x} + 0\mathbf{x}] + (-0\mathbf{x}) = (0 + 0)\mathbf{x} + (-0\mathbf{x}) \\
&= 0\mathbf{x} + (-0\mathbf{x}) = \mathbf{0}.
\end{aligned}
$$

Note also that the element $-x$ in V4 is unique. In fact if $x + y = 0$, then

$$y = y + 0 = y + [x + (-x)] = [y + x] + (-x)$$
$$= [x + y] + (-x) = 0 + (-x) = (-x) + 0 = -x.$$

This implies that $(-1)x = -x$, since

$$x + (-1)x = [1 + (-1)]x = 0x = 0.$$

A non-empty subset Y of a (complex) vector space X is called a *subspace* of X if it is closed with respect to the operations of addition and scalar multiplication. This means that if $x, y \in Y$ and $a \in \mathbb{C}$, then $x + y$ and ay are in Y. If so, then Y itself is a vector space over $\mathbb{C}$, with the operations inherited from X.

Examples

1. Any vector space is a subspace of itself.
2. The set $\{0\}$ is a subspace.
3. $\{x \mid x_n = 0\}$ is a subspace of $\mathbb{R}^n$.
4. In the previous set of examples, the space X in example 7 is a subspace of $F(\mathbb{R}; \mathbb{R})$.
5. Let X again be the space of polynomials with real coefficients. For each $n = 0, 1, 2, \ldots$, let X_n be the subset of X consisting of polynomials of degree $\leq n$. Then each X_n is a subspace of X. For $m \leq n$, X_m is a subspace of X_n.

Suppose $x_1, x_2, \ldots, x_n$ are elements of the vector space X. A *linear combination* of these vectors is any vector x of the form

$$x = a_1 x_1 + a_2 x_2 + \cdots + a_n x_n,$$

where $a_1, a_2, \ldots, a_n$ are scalars.

Proposition 7.1. *Let S be a nonempty subset of the vector space X, and let Y be the set of all linear combinations of elements of S. Then Y is a subspace of X and $Y \supset S$.*

If Z is any other subspace of X which contains the set S, then $Z \supset Y$.

Proof. If $x, y \in Y$, then by definition they can be expressed as finite sums $x = \sum a_j x_j, y = \sum b_j y_j$, where each $x_j \in S$ and each $y_j \in S$. Then $ax = \sum (aa_j)x_j$ is a linear combination of the x_j's, and $x + y$ is a linear combination of the x_j's and the y_j's. If $x \in S$, then $x = 1x \in Y$. Thus Y is a subspace containing S.

Suppose Z is another subspace of X containing S. Suppose $x \in Y$. Then for some $x_1, x_2, \ldots, x_n \in S$, $x = \sum a_j x_j$. Since Z is a subspace and $x_1, x_2, \ldots, x_n \in Z$, we have $a_1 x_1, a_2 x_2, \ldots, a_n x_n \in Z$. Moreover, $a_1 x_1 + a_2 x_2 \in Z$, so $(a_1 x_1 + a_2 x_2) + a_3 x_3 \in Z$. Continuing we eventually find that $x \in Z$. ☐

We can paraphrase Proposition 7.1 by saying that any subset S of a vector space X is contained in a unique smallest subspace Y. This subspace is called the *span* of S. We write $\mathbf{Y}$ = span (S). The set S is said to *span* $\mathbf{Y}$. Note that if S is empty, the span is the subspace $\{\mathbf{0}\}$.

Examples

Let $\mathbf{X}$ be the space of all polynomials with real coefficients. Let f_m be the polynomial defined by $f_m(x) = x^m$. Then span $\{f_0, f_1, \ldots, f_n\}$ is the subspace $\mathbf{X}_n$ of polynomials of degree $\leq n$.

A linear combination $a_1\mathbf{x}_1 + a_2\mathbf{x}_2 + \cdots + a_n\mathbf{x}_n$ of the vectors $\mathbf{x}_1, \mathbf{x}_2, \ldots,$ $\mathbf{x}_n$ is said to be *nontrivial* if at least one of the coefficients $a_1, a_2, \ldots, a_n$ is not zero. The vectors $\mathbf{x}_1, \mathbf{x}_2, \ldots, \mathbf{x}_n$ are said to be *linearly dependent* if some nontrivial linear combination of them is the zero vector. Otherwise they are said to be *linearly independent*. More generally, an arbitrary (possibly infinite) subset S is said to be linearly dependent if some nontrivial linear combination of finitely many distinct elements of S is the zero vector; otherwise S is said to be linearly independent. (Note that with this definition, the empty set is linearly independent.)

Lemma 7.2. *Vectors* $\mathbf{x}_1, \mathbf{x}_2, \ldots, \mathbf{x}_n$ *in* $\mathbf{X}$, $n \geq 2$, *are linearly dependent if and only if some* $\mathbf{x}_j$ *is a linear combination of the others.*

Proof. If $\mathbf{x}_1, \mathbf{x}_2, \ldots, \mathbf{x}_n$ are linearly dependent, there are scalars $a_1,$ $a_2, \ldots, a_n$, not all 0, such that $\sum a_j\mathbf{x}_j = 0$. Renumbering, we may suppose $a_1 \neq 0$. Then $\mathbf{x}_1 = \sum_{j=2}^{n} (-a_1^{-1}a_j)\mathbf{x}_j$.

Conversely, suppose $\mathbf{x}_1$, say, is a linear combination $\sum_{j=2}^{n} b_j\mathbf{x}_j$. Letting $a_1 = 1$, and $a_j = -b_j$ for $j \geq 2$, we have $\sum a_j\mathbf{x}_j = 0$. $\square$

The vector space $\mathbf{X}$ is said to be *finite dimensional* if there is a finite subset which spans $\mathbf{X}$. Otherwise, $\mathbf{X}$ is said to be *infinite dimensional*. A *basis* of a (finite-dimensional) space $\mathbf{X}$ is an ordered finite subset $(\mathbf{x}_1, \mathbf{x}_2, \ldots, \mathbf{x}_n)$ which is linearly independent and spans $\mathbf{X}$.

Examples

1. $\mathbb{R}^n$ has basis vectors $(\mathbf{e}_1, \mathbf{e}_2, \ldots, \mathbf{e}_n)$, where $\mathbf{e}_1 = (1, 0, 0, \ldots, 0)$, $\mathbf{e}_2 = (0, 1, 0, \ldots, 0), \ldots, \mathbf{e}_n = (0, 0, \ldots, 0, 1)$. This is called the *standard basis* in $\mathbb{R}^n$.
2. The set consisting of the single vector 1 is a basis for $\mathbb{C}$ as a *complex* vector space, but not as a *real* vector space. The set $(1, i)$ is a basis for $\mathbb{C}$ as a real vector space, but is linearly dependent if $\mathbb{C}$ is considered as a complex vector space.

Theorem 7.3. *A finite-dimensional vector space* $\mathbf{X}$ *has a basis. Any two bases of* $\mathbf{X}$ *have the same number of elements.*

Proof. Let $\{\mathbf{x}_1, \mathbf{x}_2, \ldots, \mathbf{x}_n\}$ span $\mathbf{X}$. If these vectors are linearly independent then we may order this set in any way and have a basis. Otherwise if $n \geq 2$ we may use Lemma 7.2 and renumber, so that x_n is a linear combination $\sum_{j=1}^{n-1} a_j x_j$. Since span $\{x_1, x_2, \ldots, x_n\} = X$, any $x \in X$ is a linear combination

$$x = \sum_{j=1}^{n} b_j x_j = \sum_{j=1}^{n-1} b_j x_j + b_n \left(\sum_{j=1}^{n-1} a_j x_j \right)$$
$$= \sum_{j=1}^{n-1} (b_j + b_n a_j) x_j.$$

Thus span $\{x_1, x_2, \ldots, x_{n-1}\} = X$. If these vectors are not linearly independent, we may renumber and argue as before to show that

$$\text{span} \{x_1, x_2, \ldots, x_{n-2}\} = X.$$

Eventually we reach a linearly independent subset which spans X, and thus get a basis, or else we reach a linearly dependent set $\{x_1\}$ spanning X and consisting of one element. This implies $x_1 = 0$, so $X = \{0\}$, and the empty set is the basis.

Now suppose $(x_1, x_2, \ldots, x_n)$ and $(y_1, y_2, \ldots, y_m)$ are bases of X, and suppose $m \leq n$. If $n = 0$, then $m = 0$. Otherwise $x_1 \neq 0$. The y_j's span X, so $x_1 = \sum a_j y_j$. Renumbering, we may assume $a_1 \neq 0$. Then

$$y_1 = a_1^{-1} x_1 - \sum_{j=2}^{m} a_1^{-1} a_j y_j.$$

Thus y_1 is a linear combination of $x_1, y_2, \ldots, y_m$. It follows easily that span $\{x_1, y_2, \ldots, y_m\} \supset$ span $\{y_1, y_2, \ldots, y_m\} = X$. If $m = 1$ this shows that span $\{x_1\} = X$, and the linear independence of the x_j's then implies $n = 1$. Otherwise $x_2 = b x_1 + \sum_{j=2}^{m} b_j y_j$. The independence of x_1 and x_2 implies some $b_j \neq 0$. Renumbering, we assume $b_2 \neq 0$. Then

$$y_2 = b_2^{-1} \left(x_2 - b x_1 - \sum_{j=3}^{m} b_j y_j \right).$$

This implies that

$$\text{span} \{x_1, x_2, y_3, \ldots, y_m\} \supset \text{span} \{x_1, y_2, \ldots, y_m\} = X.$$

Continuing in this way, we see that after the y_j's are suitably renumbered, each set $\{x_1, x_2, \ldots, x_k, y_{k+1}, \ldots, y_m\}$ spans X, $k \leq m$. In particular, taking $k = m$ we have that $\{x_1, x_2, \ldots, x_m\}$ spans X. Since the x_j's were assumed linearly independent, we must have $n \leq m$. Thus $n = m$. $\square$

If X has a basis with n elements, $n = 0, 1, 2, \ldots$, then any basis has n elements. The number n is called the *dimension* of n. We write $n = \dim X$.

The argument used to prove Theorem 7.3 proves somewhat more.

Theorem 7.4. *Suppose X is a finite-dimensional vector space with dimension n. Any subset of X which spans X has at least n elements. Any subset of X which is linearly independent has at most n elements. An ordered subset of n elements which either spans X or is linearly independent is a basis.*

Suppose $(x_1, x_2, \ldots, x_n)$ is a basis of X. Then any $x \in X$ can be written as a linear combination $x = \sum a_j x_j$. The scalars $a_1, a_2, \ldots, a_n$ are *unique*; in fact if $x = \sum b_j x_j$, then

$$0 = x - x = \sum (a_j - b_j) x_j.$$

Since the x_j's are linearly independent, each $a_j - b_j = 0$, i.e., $a_j = b_j$. Thus the equation $x = \sum a_j x_j$ associates to each $x \in X$ a unique ordered n-tuple $(a_1, a_2, \ldots, a_n)$ of scalars, called the *coordinates of x with respect to the basis* $(x_1, \ldots, x_n)$. Note that if x and y correspond respectively to $(a_1, a_2, \ldots, a_n)$ and $(b_1, b_2, \ldots, b_n)$, then ax corresponds to $(aa_1, aa_2, \ldots, aa_n)$ and $x + y$ corresponds to $(a_1 + b_1, a_2 + b_2, \ldots, a_n + b_n)$. In other words, the basis $(x_1, x_2, \ldots, x_n)$ gives rise to a function from X onto $\mathbb{R}^n$ or $\mathbb{C}^n$ which preserves the vector operations.

Suppose X and Y are vector spaces, either both real or both complex. A function $T: X \rightarrow Y$ is said to be *linear* if for all vectors $x, x' \in X$ and all scalars a,

$$T(ax) = aT(x), \quad T(x + y) = T(x) + T(y).$$

A linear function is often called a *linear operator* or a *linear transformation*. A linear function $T: X \rightarrow \mathbb{R}$ (for X a real vector space) or $T: X \rightarrow \mathbb{C}$ (for X a complex vector space) is called a *linear functional*.

Examples

1. Suppose X is a real vector space and $(x_1, x_2, \ldots, x_n)$ a basis. Let $T(\sum a_j x_j) = (a_1, a_2, \ldots, a_n)$. Then $T: X \rightarrow \mathbb{R}^n$ is a linear transformation.

2. Let $T(z) = z^*$, $z \in \mathbb{C}$. Then T is a linear transformation of $\mathbb{C}$ into itself if $\mathbb{C}$ is considered as a real vector space, but is *not* linear if $\mathbb{C}$ is considered as a complex vector space.

3. Let $f_j: \mathbb{R}^n \rightarrow \mathbb{R}$ be defined by $f_j(x_1, x_2, \ldots, x_n) = x_j$. Then f_j is a linear functional.

4. Let X be the space of polynomials with real coefficients. The two functions S, T defined below are linear transformations from X to itself. If $f(x) = \sum_{j=0}^{n} a_j x^j$, then

$$S(f)(x) = \sum_{j=0}^{n} (j + 1)^{-1} a_j x^{j+1},$$

$$T(f)(x) = \sum_{j=1}^{n} j a_j x^{j-1}.$$

Note that $T(S(f)) = f$, while $S(T(f)) = f$ if and only if $a_0 = 0$.

Exercises

1. If the linearly independent finite set $\{x_1, x_2, \ldots, x_n\}$ does not span X, then there is a vector $x_{n+1} \in X$ such that $\{x_1, x_2, \ldots, x_n, x_{n+1}\}$ is linearly independent.

2. If X is finite-dimensional and $x_1, x_2, \ldots, x_n$ are linearly independent, then there is a basis of X containing the vectors $x_1, x_2, \ldots, x_n$.

3. If X is a finite-dimensional vector space and $Y \subset X$ is a subspace, then Y is finite-dimensional. Moreover, dim $Y \leq$ dim X, and dim $Y <$ dim X unless $Y = X$.

4. If Y and Z are subspaces of the finite-dimensional vector space X and $Y \cap Z = \{0\}$, then dim $Y +$ dim $Z \leq$ dim X.

5. Suppose Y and Z are subspaces of the finite-dimensional vector space X, and suppose span $(Y \cup Z) = X$. Then dim $Y +$ dim $Z \geq$ dim X.

6. If Y is a subspace of the finite dimensional vector space X, then there is a subspace Z with the properties $Y \cap Z = \{0\}$; dim $Y +$ dim $Z =$ dim X; any vector $x \in X$ can be expressed uniquely in the form $x = y + z$, where $y \in Y$, $z \in Z$. (Such subspaces are said to be *complementary*.)

7. Prove Theorem 7.4.

8. The polynomials $f_m(x) = x^m$, $0 \leq m \leq n$, are a basis for the vector space of polynomials of degree $\leq n$.

9. The vector space of all polynomials is infinite dimensional.

10. If S is a non-empty set, the vector space $F(S; \mathbb{R})$ of functions from S to $\mathbb{R}$ is finite dimensional if and only if S is finite.

11. If X and Y are vector spaces and $T: X \to Y$ is a linear transformation, then the sets

$$N(T) = \{x \mid x \in X, T(x) = 0\}$$
$$R(T) = \{T(x) \mid x \in X\}$$

are subspaces of X and Y, respectively. (They are called the *null space* or *kernel* of T, and the *range* of T, respectively.) T is 1-1 if and only if $N(T) = \{0\}$.

12. If X is finite dimensional, the subspaces $N(T)$ and $R(T)$ in problem 11 satisfy dim $N(T) +$ dim $R(T) =$ dim X. In particular, if dim $Y =$ dim X, then T is 1-1 if and only if it is onto. (Hint: choose a basis for $N(T)$ and use problem 2 to extend to a basis for X. Then the images under T of the basis elements not in $N(T)$ are a basis for $R(T)$.)

Chapter 2

Continuous Functions

§1. Continuity, uniform continuity, and compactness

Suppose (S, d) and (S', d') are metric spaces. A function $f: S \to S'$ is said to be *continuous at the point* $x \in S$ if for each $\varepsilon > 0$ there is a $\delta > 0$ such that

$$d'(f(x), f(y)) < \varepsilon \qquad \text{if } d(x, y) < \delta.$$

In particular, if S and S' are subsets of $\mathbb{R}$ or of $\mathbb{C}$ (with the usual metrics) then the condition is

$$|f(y) - f(x)| < \varepsilon \qquad \text{if } |y - x| < \delta.$$

(This definition is equivalent to the following one, given in terms of convergence of sequences: f is continuous at x if $f(x_n) \to f(x)$ whenever $(x_n)_{n=1}^{\infty}$ is a sequence in S which converges to x. The equivalence is left as an exercise.)

Recall that we can add, multiply, and take scalar multiples of functions with values in $\mathbb{C}$ (or $\mathbb{R}$): if $f, g: S \to \mathbb{C}$ and $a \in \mathbb{C}$, $x \in S$, then

$$(f + g)(x) = f(x) + g(x),$$
$$(af)(x) = af(x),$$
$$(fg)(x) = f(x)g(x),$$
$$(f/g)(x) = f(x)/g(x) \qquad \text{if } g(x) \neq 0.$$

Proposition 1.1. *Suppose (S, d) is any metric space, and suppose $f, g: S \to \mathbb{C}$ are functions which are continuous at x. Then $f + g$, af, and fg are continuous at x. If $g(x) \neq 0$, then f/g is defined in a ball $B_r(x)$ and is continuous at x.*

Proof. Continuity of $f + g$ and af at x follow from the definition of continuity and the inequalities

$$|(f + g)(y) - (f + g)(x)| = |f(y) - f(x) + g(y) - g(x)|$$
$$\leq |f(y) - f(x)| + |g(y) - g(x)|,$$
$$|(af)(y) - (af)(x)| = |a| \, |f(y) - f(x)|.$$

To show continuity of fg at x we choose $\delta_1 > 0$ so small that if $d(y, x) < \delta_1$ then $|f(y) - f(x)| < 1$. Let M be the larger of $|g(x)|$ and $|f(x)| + 1$. Given $\varepsilon > 0$, choose $\delta > 0$ so small that

$$|f(y) - f(x)| < \varepsilon/2M, \qquad |g(y) - g(x)| < \varepsilon/2M$$

34

if $d(y, x) < \delta$. Then $d(y, x) < \delta$ implies

$$\begin{aligned}|(fg)(y) - (fg)(x)| &= |f(y)g(y) - f(x)g(x)| \\ &= |f(y)g(y) - f(y)g(x) + f(y)g(x) - f(x)g(x)| \\ &\le |f(y)| \, |g(y) - g(x)| + |f(y) - f(x)| \, |g(x)| \\ &\le |f(y)| \cdot \varepsilon/2M + M \cdot \varepsilon/2M \\ &= |f(y)| \cdot \varepsilon/2M + \varepsilon/2. \end{aligned}$$

But also

$$\begin{aligned}|f(y)| = |f(y) - f(x) + f(x)| &\le |f(y) - f(x)| + |f(x)| \\ &< 1 + |f(x)| \le M, \end{aligned}$$

so $|(fg)(y) - (fg)(x)| < \varepsilon$.

Finally, suppose $g(x) \ne 0$. Choose $r > 0$ so that $|g(y) - g(x)| < \frac{1}{2}|g(x)|$ if $d(y, x) < r$. Then if $d(y, x) < r$ we have

$$\begin{aligned}|g(x)| = |g(y) + g(x) - g(y)| &\le |g(y)| + \tfrac{1}{2}|g(x)|, \end{aligned}$$

so $|g(y)| \ge \frac{1}{2}|g(x)| > 0$. Thus $1/g$ is defined on $B_r(x)$. Since the product of functions continuous at x is continuous at x, we only need show that $1/g$ is continuous at x. But if $y \in B_r(x)$, then

$$\begin{aligned}|1/g(y) - 1/g(x)| = |g(y) - g(x)|/|g(y)| \, |g(x)| \\ \le K|g(y) - g(x)|, \end{aligned}$$

where $K = 2/|g(x)|^2$. Since g is continuous at x, it follows that $1/g$ is. $\quad\square$

A function $f: S \to S'$ is said to be *continuous* if it is continuous at each point of S.

The following is an immediate consequence of Proposition 1.1.

Corollary 1.2. *Suppose $f, g: S \to \mathbb{C}$ are continuous. Then $f + g$, af, and fg are also continuous. If $g(x) \ne 0$, all x, then f/g is continuous.*

A function $f: S \to S'$ is said to be *uniformly continuous* if for each $\varepsilon > 0$ there is a $\delta > 0$ such that

$$d'(f(x), f(y)) < \varepsilon \qquad \text{if } d(x, y) < \delta.$$

In particular, if $S, S' \subset \mathbb{C}$, then this condition reads

$$|f(y) - f(x)| < \varepsilon \qquad \text{if } |y - x| < \delta.$$

The distinction between continuity and uniform continuity is important. If f is continuous, then for *each* x and each $\varepsilon > 0$ there is a $\delta > 0$ such that the above condition holds; however, δ may depend on x. As an example, let $S = S' = \mathbb{R}, f(x) = x^2$. Then $|f(y) - f(x)| = |y^2 - x^2| = |y + x| \, |y - x|$. If $|x|$ is very large, then $|y - x|$ must be very small for $|f(y) - f(x)|$ to be less than 1. Thus this function is not uniformly continuous. (However it is clear that any uniformly continuous function is continuous.)

In one important case, continuity *does* imply uniform continuity.

Theorem 1.3. *Suppose* (S, d) *and* (S', d') *are metric spaces and suppose* $f: S \to S'$ *is continuous. If* S *is compact, then* f *is uniformly continuous.*

Proof. Given $\varepsilon > 0$, we know that for each $x \in S$ there is a number $\delta(x) > 0$ such that

$$d'(f(x), f(y)) < \tfrac{1}{2}\varepsilon \qquad \text{if } d(x, y) < 2\delta(x).$$

Let $N(x) = B_{\delta(x)}(x)$. By the definition of compactness, there are points $x_1, x_2, \ldots, x_n \in S$ such that $S \subset \bigcup N(x_j)$. Let $\delta = \min \{\delta(x_1), \delta(x_2), \ldots, \delta(x_n)\}$, and suppose $d(x, y) < \delta$. There is some x_i such that $x \in N(x_i)$. Then

$$d(x_i, x) < \delta(x_i) < 2\delta(x_i),$$
$$d(x_i, y) < d(x_i, x) + d(x, y) < \delta(x_i) + \delta \le 2\delta(x_i),$$

so

$$d'(f(x), f(y)) \le d'(f(x), f(x_i)) + d'(f(x_i), f(y))$$
$$\le \tfrac{1}{2}\varepsilon + \tfrac{1}{2}\varepsilon = \varepsilon. \qquad \square$$

There are other pleasant properties of continuous functions on compact sets. A function $f: S \to \mathbb{C}$ is said to be *bounded* if $f(S)$ is a bounded set in $\mathbb{C}$, i.e., there is an $M \ge 0$ such that

$$|f(x)| \le M, \qquad \text{all } x \in S.$$

Theorem 1.4. *Suppose* (S, d) *is a compact metric space and suppose* $f: S \to \mathbb{C}$ *is continuous. Then* f *is bounded and there is a point* $x_0 \in S$ *such that*

$$|f(x_0)| = \sup \{|f(x)| \mid x \in S\}.$$

If $f(S) \subset \mathbb{R}$, *then there are* $x_+, x_- \in S$ *such that*

$$f(x_+) = \sup \{f(x) \mid x \in S\},$$
$$f(x_-) = \inf \{f(x) \mid x \in S\}.$$

Proof. For each $x \in S$, there is a number $\delta(x) > 0$ such that $|f(y) - f(x)| < 1$ if $y \in B_{\delta(x)}(x) = N(x)$. Choose $x_1, \ldots, x_n$ such that $S \subset \bigcup N(x_j)$. If $x \in S$ then $x \in N(x_i)$ for some i, and

$$|f(x)| \le |f(x_i)| + |f(x) - f(x_i)| < |f(x_i)| + 1.$$

Thus we can take $M = 1 + \max \{|f(x_1)|, \ldots, |f(x_n)|\}$ and we have shown that f is bounded.

Let $a = \sup \{|f(x)| \mid x \in S\}$, and suppose $|f(x)| < a$, all $x \in S$. Then for each $x \in S$ there are numbers $a(x) < a$ and $\varepsilon(x) > 0$ such that $|f(y)| \le a(x)$ if $y \in B_{\varepsilon(x)}(x) = M(x)$. Choose $y_1, \ldots, y_m$ such that $S \subset \bigcup M(y_i)$, and let $a_1 = \max \{a(y_1), \ldots, a(y_m)\} < a$. If $x \in S$ then $x \in M(y_i)$ for some i, so

$$|f(x)| < a(y_i) \le a_1 < a.$$

This contradicts the assumption that $a = \sup \{|f(x)| \mid x \in S\}$. Thus there must be a point x_0 with $|f(x)| = a$.

The proof of the existence of x_+ and x_- when f is real-valued is similar, and we omit it. $\square$

Both theorems above apply in particular to continuous functions defined on a closed bounded interval $[a, b] \subset \mathbb{R}$. We need one further fact about such functions when real-valued: they skip no values.

Theorem 1.5. (Intermediate Value Theorem). *Suppose $f: [a, b] \to \mathbb{R}$ is continuous. Suppose either*

$$f(a) \leq c \leq f(b) \quad or \quad f(b) \leq c \leq f(a).$$

Then there is a point $x_0 \in [a, b]$ such that $f(x_0) = c$.

Proof. We consider only the case $f(a) \leq c \leq f(b)$. Consider the two intervals $[a, \frac{1}{2}(a + b)]$, $[\frac{1}{2}(a + b), b]$. For at least one of these, c lies between the values of f at the endpoints; denote this subinterval by $[a_1, b_1]$. Thus $f(a_1) \leq c \leq f(b_1)$. Continuing in this way we get a sequence of intervals $[a_n, b_n]$ with $[a_{n+1}, b_{n+1}] \subset [a_n, b_n]$, $b_{n+1} - a_{n+1} = \frac{1}{2}(b_n - a_n)$, and $f(a_n) \leq c \leq f(b_n)$. Then there is $x_0 \in [a, b]$ such that $a_n \to x_0$, $b_n \to x_0$. Thus

$$f(x_0) = \lim f(a_n) \leq c.$$
$$f(x_0) = \lim f(b_n) \geq c. \qquad \qquad \square$$

Exercises

1. Prove the equivalence of the two definitions of continuity at a point.

2. Use Theorem 1.5 to give another proof of the existence of $\sqrt{2}$. Prove that any positive real number has a positive nth root, $n = 1, 2, \ldots$.

3. Suppose $f: S \to S'$, where (S, d) and (S', d') are metric spaces. Prove that the following are equivalent:
   (a) f is continuous;
   (b) for each open set $A' \subset S'$, $f^{-1}(A')$ is open;
   (c) for each closed set $A' \subset S'$, $f^{-1}(A')$ is closed.

4. Find continuous functions $f_j: (0, 1) \to \mathbb{R}$, $j = 1, 2, 3$, such that
   f_1 is not bounded,
   f_2 is bounded but not uniformly continuous,
   f_3 is bounded but there are no points $x_+, x_- \in (0, 1)$ such that $f_3(x_+) = \sup \{f_3(x) \mid x \in (0, 1)\}$, $f_3(x_-) = \inf \{f_3(x) \mid x \in (0, 1)\}$.

5. Suppose $f: S \to S'$ is continuous and S is compact. Prove that $f(S)$ is compact.

6. Use Exercise 5 and Theorem 6.2 of Chapter 1 to give another proof of Theorem 1.4.

7. Use Exercise 3 of Chapter 1, §6 to give a third proof of Theorem 1.4. (Hint: take $(x_n)_{n=1}^{\infty} \subset S$ such that $\lim |f(x_n)| = \sup \{f(x) \mid x \in S\}$, etc.)

8. Suppose (S, d) is a metric space, $x \in S$, and $r > 0$. Show that there is a continuous function $f: S \to \mathbb{R}$ with the properties: $0 \leq f(y) \leq 1$, all $y \in S$, $f(y) = 0$ if $y \notin B_r(x)$, $f(x) = 1$. (Hint: take $f(y) = \max \{1 - r^{-1}d(y, x), 0\}$.)

9. Suppose (S, d) is a metric space and suppose S is *not* compact. Show that there is a continuous $f: S \to \mathbb{R}$ which is not bounded. (Hint: use Exercise 8.)

§2. Integration of complex-valued functions

A *partition* of a closed interval $[a, b] \subset \mathbb{R}$ is a finite ordered set of points $P = (x_0, x_1, \ldots, x_n)$ with

$$a = x_0 < x_1 < \cdots < x_n = b.$$

The *mesh* of the partition P is the maximum length of the subintervals $[x_{i-1}, x_i]$:

$$|P| = \max \{x_i - x_{i-1} \mid i = 1, 2, \ldots, n\}.$$

If $f: [a, b] \to \mathbb{C}$ is a bounded function and $P = (x_0, x_1, \ldots, x_n)$ is a partition of $[a, b]$, then the *Riemann sum of f* associated with the partition P is the number

$$S(f; P) = \sum_{i=1}^{n} f(x_i)(x_i - x_{i-1}).$$

The function f is said to be *integrable* (in the sense of Riemann) if there is number $z \in \mathbb{C}$ such that

$$\lim_{|P| \to 0} S(f; P) = z.$$

More precisely, we mean that for any $\varepsilon > 0$ there is a $\delta > 0$ such that

$$(2.1) \qquad\qquad |S(f; P) - z| < \varepsilon \qquad \text{if } |P| < \delta.$$

If this is the case, the number z is called the *integral of f* on $[a, b]$ and denoted by

$$\int_a^b f \quad \text{or} \quad \int_a^b f(x)\, dx.$$

If $f: [a, b] \to \mathbb{C}$ is bounded, suppose $|f(x)| \leq M$, all $x \in [a, b]$. Then for any partition P of $[a, b]$,

$$|S(f; P)| \leq \sum |f(x_i)|(x_i - x_{i-1}) \leq M \sum (x_i - x_{i-1}) = M(b - a).$$

Therefore, if f is integrable,

$$(2.2) \qquad \left| \int_a^b f \right| \leq M(b - a), \qquad M = \sup \{|f(x)| \mid x \in [a, b]\}.$$

Recall that $f: [a, b] \to \mathbb{C}$ is a sum $f = g + ih$ where g and h are real-valued functions. The functions g and h are called the *real* and *imaginary parts* of f and are defined by

$$g(x) = \operatorname{Re}(f(x)), \qquad h(x) = \operatorname{Im}(f(x)), \qquad x \in [a, b].$$

We denote g by $\operatorname{Re} f$ and h by $\operatorname{Im} f$.

Proposition 2.1. *A bounded function $f: [a, b] \to \mathbb{C}$ is integrable if and only if the real and imaginary parts of f are integrable. If so,*

$$\int_a^b f = \int_a^b \operatorname{Re} f + i \int_a^b \operatorname{Im} f.$$

Proof.   Recall that if $z = x + iy$, $x, y \in \mathbb{R}$, then

(2.3) $\tfrac{1}{2}|x| + \tfrac{1}{2}|y| \le |z| \le |x| + |y|.$

If P is any partition of $[a, b]$, then

$$S(f; P) = S(\mathrm{Re}\, f; P) + iS(\mathrm{Im}\, f; P),$$

and $S(\mathrm{Re}\, f; P)$, $S(\mathrm{Im}\, f; P)$ are real. Let $z = x + iy$, x, y real, and apply
(2.3) to $S(f; P) - z$. We get

$$\tfrac{1}{2}|S(\mathrm{Re}\, f; P) - x| + \tfrac{1}{2}|S(\mathrm{Im}\, f; P) - y|$$
$$\le |S(f; P) - z| \le |S(\mathrm{Re}\, f; P) - x| + |S(\mathrm{Im}\, f; P) - y|.$$

Thus $S(f; P) \to z$ as $|P| \to 0$ if and only if $S(\mathrm{Re}\, f; P) \to x$ and $S(\mathrm{Im}\, f; P) \to y$ as $|P| \to 0$.   □

Proposition 2.2.   *Suppose $f: [a, b] \to \mathbb{C}$ and $g: [a, b] \to \mathbb{C}$ are bounded integrable functions, and suppose $c \in \mathbb{C}$. Then $f + g$ and cf are integrable, and*

$$\int_a^b (f + g) = \int_a^b f + \int_a^b g, \qquad \int_a^b cf = c \int_a^b f.$$

Proof.   For any partition P of $[a, b]$,

$$S(f + g; P) = S(f; P) + S(g; P), \qquad S(cf; P) = cS(f; P).$$

The conclusions follow easily from these identities and the definition.   □

Neither of these propositions identifies any integrable functions. We shall see shortly that continuous functions are integrable. The following criterion is useful in that connection.

Proposition 2.3.   *A bounded function $f: [a, b] \to \mathbb{C}$ is integrable if and only if for each $\varepsilon > 0$ there is a $\delta > 0$ such that*

(2.4) $|S(f; P) - S(f; Q)| < \varepsilon$ *if $|P|, |Q| < \delta.$*

Proof.   Suppose f is integrable, and let $z = \int_a^b f$. For any $\varepsilon > 0$ there is a $\delta > 0$ such that $S(f; P)$ is in the disc of radius $\tfrac{1}{2}\varepsilon$ around z if $|P| < \delta$. Then $|P| < \delta, |Q| < \delta$ implies $S(f; P), S(f; Q)$ are at distance $< \varepsilon$.

Conversely, suppose for each $\varepsilon > 0$ there is a $\delta > 0$ such that (2.4) holds. Take partitions P_n with $|P_n| < 1/n$, $n = 1, 2, 3, \ldots$, and let $z_n = S(f; P_n)$. It follows from our assumption that $(z_n)_{n=1}^\infty$ is a Cauchy sequence. Let z be its limit. If n is large, $|P_n|$ is small and $S(f; P_n)$ is close to z, and if $|Q|$ is small, $S(f; Q)$ is close to $S(f; P_n)$. Thus $S(f; Q) \to z$ as $|Q| \to 0$.   □

Theorem 2.4.   *If $f: [a, b] \to \mathbb{C}$ is continuous, it is integrable.*

Proof.   We know by §1 that f is bounded and uniformly continuous. Given $\varepsilon > 0$, choose $\delta > 0$ so that $|f(x) - f(y)| < \varepsilon$ if $|x - y| < \delta$. Suppose P, Q are partitions of $[a, b]$ with $|P| < \delta$, $|Q| < \delta$. Suppose $P = (x_0, x_1, \ldots, x_n)$. Let P' be a partition which includes all points of P and of Q, $P' = (y_0, y_1, \ldots, y_m)$. We examine one summand of $S(f; P)$. Suppose

$$x_{i-1} = y_{j-1} < y_j < \cdots < y_k = x_i.$$

Then

$$\left| f(x_i)(x_i - x_{i-1}) - \sum_{l=j}^{k} f(y_l)(y_l - y_{l-1}) \right|$$

$$= \left| \sum_{l=j}^{k} (f(x_i) - f(y_l))(y_l - y_{l-1}) \right|$$

$$\le \varepsilon \sum_{l=j}^{k} (y_l - y_{l-1}) = \varepsilon(x_i - x_{i-1}),$$

since each $|x_i - y_l| < \delta$. Adding, we get

$$|S(f; P) - S(f; P')| < \varepsilon(b - a).$$

Similarly,

$$|S(f; Q) - S(f; P')| < \varepsilon(b - a),$$

so

$$|S(f; P) - S(f; Q)| < 2\varepsilon(b - a).$$

By Proposition 2.3, f is integrable. □

We now want to consider the effect of integrating over subintervals.

Proposition 2.5. *Suppose $a < b < c$ and $f: [a, c] \to \mathbb{C}$ is bounded. Then f is integrable if and only if it is integrable as a function on $[a, b]$ and on $[b, c]$. If so, then*

$$\int_a^c f = \int_a^b f + \int_b^c f.$$

Proof. Suppose f is integrable on $[a, b]$ and on $[b, c]$. Given $\varepsilon > 0$, choose $\delta > 0$ so that

$$(2.5) \qquad \left| S(f; P) - \int_a^b f \right| < \frac{1}{2}\varepsilon, \qquad \left| S(f; Q) - \int_b^c f \right| < \frac{1}{2}\varepsilon$$

if P, Q are partitions of $[a, b]$, $[b, c]$ respectively, $|P| < \delta$, $|Q| < \delta$. Suppose P' is a partition of $[a, c]$, $|P'| < \delta$. If b is a point of P', then P' determines partitions P of $[a, b]$ and Q of $[b, c]$, $|P| < \delta$, $|Q| < \delta$. It follows from (2.5) that

$$(2.6) \qquad \left| S(f; P') - \int_a^b f - \int_b^c f \right| < \varepsilon.$$

If b is not a point of P', let P'' be the partition obtained by adjoining b. Then (2.6) holds with P'' in place of P'. Suppose $|f(x)| \le M$, all $x \in [a, b]$. The sums $S(f; P')$ and $S(f; P'')$ differ only in terms corresponding to the subinterval determined by P' which contains b. It is easy to check, then, that

$$|S(f; P') - S(f; P'')| < 2\delta M.$$

Thus

$$\left| S(f; P') - \int_a^b f - \int_b^c f \right| < \varepsilon + 2\delta M$$

if $|P'| < \delta$. It follows that f is integrable with integral $\int_a^b f + \int_b^c f$.

Conversely, suppose f is *not* integrable on $[a, b]$ or $[b, c]$; say it is not integrable on $[a, b]$. Then there is an $\varepsilon > 0$ such that for any $\delta > 0$ there are partitions P_1, P_2 of $[a, b]$ with $|P_1| < \delta$, $|P_2| < \delta$, but

$$|S(f; P_1) - S(f; P_2)| > \varepsilon.$$

Let Q be a partition of $[b, c]$ with $|Q| < \delta$, and let P_1', P_2' be the partitions of $[a, c]$ containing all points of Q and of P_1, P_2 respectively. Then $|P_1'| < \delta$, $|P_2'| < \delta$, and

$$|S(f; P_1') - S(f; P_2')| = |S(f; P_1) - S(f; P_2)| > \varepsilon.$$

By Proposition 1.3, f is not integrable.   □

Suppose $f: [a_0, b_0] \to \mathbb{C}$ is integrable, and suppose $a, b \in [a_0, b_0]$. If $a < b$, then f is integrable on $[a, b]$. (In fact f is integrable on $[a_0, b]$, therefore on $[a, b]$, by two applications of Proposition 2.5.) If $b < a$, then f is integrable on $[b, a]$ and we *define*

$$\int_a^b f = -\int_b^a f.$$

We also *define*

$$\int_a^a f = 0.$$

Then one can easily check, case by case, that for *any* $a, b, c \in [a_0, b_0]$,

$$(2.7) \qquad \int_a^c f = \int_a^b f + \int_b^c f.$$

It is convenient to extend the notion of the integral to certain unbounded functions and to certain functions on unbounded intervals; such integrals are called *improper integrals*. We give two examples, and leave the remaining cases to the reader.

Suppose $f: (a_0, b] \to \mathbb{C}$ is bounded and integrable on each subinterval $[a, b]$, $a_0 < a < b$. We set

$$(2.8) \qquad \int_{a_0}^b f = \lim_{a \to a_0} \int_a^b f$$

if the limit exists.

Suppose $f: [a, \infty) \to \mathbb{C}$ is bounded and integrable on each subinterval $[a, b]$, $a < b < \infty$. We set

$$(2.9) \qquad \int_a^\infty f = \lim_{b \to \infty} \int_a^b f$$

if the limit exists; this means that there is a $z \in \mathbb{C}$ such that for each $\epsilon > 0$

$$\left| \int_a^b f - z \right| < \varepsilon \qquad \text{if } b \geq b(\varepsilon).$$

Exercises

1. Let $f: [0, 1] \to \mathbb{R}$ with $f(x) = 0$ if x is irrational, $f(x) = 1$ if x is rational. Show that f is not integrable.

2. Let $f: [0, 1] \to \mathbb{R}$ be defined by: $f(0) = 0$, $f(x) = \sin(1/x)$ if $x \neq 0$. Sketch the graph. Show that f is integrable.

3. Suppose $f, g: [a, b] \to \mathbb{C}$ are bounded, f is integrable, and $g(x) = f(x)$ except on a finite set of points in $[a, b]$. Then g is integrable and $\int_a^b g = \int_a^b f$.

4. Suppose $f: [a, b] \to \mathbb{C}$ is bounded and is continuous except at some finite set of points in $[a, b]$. Show that f is integrable.

5. Suppose $f: [a, b] \to \mathbb{C}$ is continuous and $f(x) \geq 0$, all $x \in [a, b]$. Show that $\int_a^b f = 0$ implies $f(x) = 0$, all $x \in [a, b]$.

6. Suppose $f: [a, b] \to \mathbb{C}$ is bounded, integrable, and real-valued. Suppose

$$\left| \int_a^b f \right| = M(b - a), \quad \text{where} \quad M = \sup\{|f(x)| \mid x \in [a, b]\}.$$

Show that f is constant.

7. Do Exercise 6 without the assumption that f is real-valued.

8. Let $f: [0, 1] \to \mathbb{C}$ be defined by: $f(x) = 0$ if $x = 0$ or x is irrational, $f(x) = 1/q$ if $x = p/q$, p, q relatively prime positive integers. Show that f is continuous at x if and only if x is zero or irrational. Show that f is integrable and $\int_0^1 f = 0$.

§3. Differentiation of complex-valued functions

Suppose (a, b) is an open interval in $\mathbb{R}$ and that $f: (a, b) \to \mathbb{C}$. As in the case of a real-valued function, we say that the function f is *differentiable* at the point $x \in (a, b)$ if the limit

(3.1) $$\lim_{y \to x} \frac{f(y) - f(x)}{y - x}$$

exists. More precisely, this means that there is a number $z \in \mathbb{C}$ such that for any $\varepsilon > 0$, there is a $\delta > 0$ with

(3.2) $$|(f(y) - f(x))(y - x)^{-1} - z| < \varepsilon \qquad \text{if } 0 < |y - x| < \delta.$$

If so, the (unique) number z is called the *derivative* of f at x and denoted variously by

$$f'(x), \quad Df(x), \quad \text{or} \quad \frac{df}{dx}(x).$$

Proposition 3.1. *If $f: (a, b) \to \mathbb{C}$ is differentiable at $x \in (a, b)$ then f is continuous at x.*

Proof. Choose $\delta > 0$ so that (3.2) holds with $z = f'(x)$ and $\varepsilon = 1$. Then when $|y - x| < \delta$ we have

$$|f(y) - f(x)| \le |f(y) - f(x) - z(y - x)| + |z(y - x)|$$
$$< (1 + |z|)|y - x|.$$

As $y \to x$, $f(y) \to f(x)$. $\square$

Proposition 3.2. *The function f: $(a, b) \to \mathbb{C}$ is differentiable at $x \in (a, b)$ if and only if the real and imaginary parts $g = \mathrm{Re}\, f$ and $h = \mathrm{Im}\, f$ are differentiable at x. If so, then*

$$f'(x) = g'(x) + ih'(x).$$

Proof. As in the proof of Proposition 2.1, the limit (3.1) exists if and only if the limits of the real and imaginary parts of this expression exist. If so, these are respectively $g'(x)$ and $h'(x)$. $\square$

Proposition 3.3. *Suppose f: $(a, b) \to \mathbb{C}$ and g: $(a, b) \to \mathbb{C}$ are differentiable at $x \in (a, b)$, and suppose $c \in \mathbb{C}$. Then the functions $f + g$, cf, and fg are differentiable at x, and*

$$(f + g)'(x) = f'(x) + g'(x),$$
$$(cf)'(x) = cf'(x),$$
$$(fg)'(x) = f'(x)g(x) + f(x)g'(x).$$

If $g(x) \ne 0$ then f/g is differentiable at x and

$$(f/g)'(x) = [f'(x)g(x) - f(x)g'(x)]g(x)^{-2}.$$

Proof. This can be proved by reducing it to the (presumed known) theorem for real-valued functions, using Proposition 3.2. An alternative is simply to repeat the proofs, which are no different in the complex case. We shall do this for the product, as an example. We have

$$(fg)(y) - (fg)(x) = f(y)g(y) - f(x)g(x)$$
$$= [f(y) - f(x)]g(y) + f(x)[g(y) - g(x)].$$

Divide by $(y - x)$ and let $y \to x$. Since $g(y) \to g(x)$, the first term converges to $f'(x)g(x)$. The second converges to $f(x)g'(x)$. $\square$

We recall the following theorem, which is only valid for real-valued functions.

Theorem 3.4. (Mean Value Theorem). *Suppose f: $[a, b] \to \mathbb{R}$ is continuous, and is differentiable at each point of (a, b). Then there is a $c \in (a, b)$ such that*

$$f'(c) = [f(b) - f(a)](b - a)^{-1}.$$

Proof. Suppose first that $f(b) = f(a)$. By Theorem 1.4 there are points c_+ and c_- in $[a, b]$ such that $f(c_+) \ge f(x)$, all $x \in [a, b]$ and $f(c_-) \le f(x)$, all $x \in [a, b]$. If c_+ and c_- are both either a or b, then f is constant and $f'(c) = 0$, all $c \in (a, b)$. Otherwise, suppose $c_+ \in (a, b)$. It follows that $[f(y) - f(c_+)] \times$ $(y - c_+)^{-1}$ is ≥ 0 if $y < c_+$ and ≤ 0 if $y > c_+$. Therefore the limit as $y \to c_+$ is zero. Similarly, if $c_- \neq a$ and $c_- \neq b$, then $f'(c_-) = 0$. Thus in this case $f'(c) = 0$ for some $c \in (a, b)$.

In the general case, let

$$g(x) = f(x) - (x - a)[f(b) - f(a)](b - a)^{-1}.$$

Then $g(a) = f(a) = g(b)$. By what we have just proved, there is a $c \in (a, b)$ such that

$$0 = g'(c) = f'(c) - [f(b) - f(a)](b - a)^{-1}. \qquad \square$$

Corollary 3.5. *Suppose f: $[a, b] \to \mathbb{C}$ is continuous, and is differentiable at each point of (a, b). If $f'(x) = 0$ for each $x \in (a, b)$, then f is constant.*

Proof. Let g, h be the real and imaginary parts of f. Then $g'(x) = h'(x) = 0$, $x \in (a, b)$. We want to show g, h constant. If $[x, y] \subset [a, b]$, Theorem 3.1 applied to g, h on $[x, y]$ implies $g(x) = g(y)$, $h(x) = h(y)$. $\quad \square$

Theorem 3.6. (Fundamental Theorem of Calculus). *Suppose f: $[a, b] \to \mathbb{C}$ is continuous and suppose $c \in [a, b]$. The function F: $[a, b] \to \mathbb{C}$ defined by*

$$F(x) = \int_c^x f$$

is differentiable at each point x of (a, b) and

$$F'(x) = f(x).$$

Proof. Let g be the constant function $g(y) = f(x)$, $y \in (a, b)$. Given $\varepsilon > 0$, choose δ so small that $|f(y) - g(y)| = |f(y) - f(x)| < \varepsilon$ if $|y - x| < \delta$. Then

(3.3)
$$F(y) - F(x) = \int_c^y f - \int_c^x f = \int_x^y f$$

$$= \int_x^y g + \int_x^y (f - g)$$

$$= f(x)(y - x) + \int_x^y (f - g).$$

If $|y - x| < \delta$, then

$$\left| \int_x^y (f - g) \right| \leq |y - x| \sup \{|f(r) - f(x)| \mid |r - x| < |y - x|\}$$
$$< \varepsilon |y - x|.$$

Thus dividing (3.3) by $(y - x)$ we get

$$|[F(y) - F(x)](y - x)^{-1} - f(x)| < \varepsilon. \qquad \square$$

Theorem 3.7. *Suppose f: $[a, b] \to \mathbb{R}$ is continuous and differentiable at each point of (a, b) and suppose $f'(x) > 0$, all $x \in (a, b)$. Then f is strictly increasing* on (a, b). *For each* $y \in [f(a), f(b)]$ *there is a unique point* $x = g(y) \in [a, b]$ *such that* $f(x) = y$. *The function* $g = f^{-1}$ *is differentiable at each point of* $(f(a), f(b))$ *and*

$$g'(y) = [f'(g(y))]^{-1}.$$

Proof. If $x, y \in [a, b]$ and $x < y$, application of Theorem 3.4 to $[x, y]$ shows that $f(x) < f(y)$. In particular, $f(a) < f(b)$. By Theorem 1.6, if $f(a) \le y \le f(b)$ there is $x \in [a, b]$ with $f(x) = y$. Since f is strictly increasing, x is unique. Letting $g = f^{-1}$ we note that g is continuous. In fact, suppose $y \in (f(a), f(b))$ and $\varepsilon > 0$. Take y', y'' such that

$$f(a) \le y' < y < y'' \le f(b)$$

and $y'' - y \le \varepsilon$, $y - y' \le \varepsilon$. Let $x' = g(y')$, $x = g(y)$, $x'' = g(y'')$. Then $x' < x < x''$. Let $\delta = \min \{x'' - x, x - x'\}$. If $|x - w| < \delta$ then $w \in (x', x'')$, so $f(w) \in (y', y'')$, so $|f(w) - f(x)| = |f(w) - y| < \varepsilon$. Continuity at $f(a)$ and $f(b)$ is proved similarly.

Finally, let $x = g(y)$, $x' = g(y')$. Then

(3.3)
$$\frac{g(y') - g(y)}{y' - y} = \frac{x' - x}{f(x') - f(x)}.$$

As $y' \to y$, we have shown that $x' \to x$. Thus (3.3) converges to $f'(x)^{-1} = [f'(g(y))]^{-1}$. $\square$

Proposition 3.8. (Chain rule). *Suppose* $g \colon (a, b) \to \mathbb{R}$ *is differentiable at* x, *and suppose* $f \colon g((a, b)) \to \mathbb{C}$ *is differentiable at* $g(x)$. *Then the composite function* $f \circ g$ *is differentiable at* x *and*

(3.4)
$$(f \circ g)'(x) = f'(g(x))g'(x).$$

Proof. We have

(3.5)
$$f \circ g(y) - f \circ g(x) = f(g(y)) - f(g(x))$$

$$= \frac{f(g(y)) - f(g(x))}{g(y) - g(x)} \cdot \frac{g(y) - g(x)}{y - x} \cdot (y - x)$$

if $g(y) \ne g(x)$. If $g'(x) \ne 0$ then $g(y) \ne g(x)$ if y is close to x and $y \ne x$. Taking the limit as $y \to x$ we get (3.4). Suppose $g'(x) = 0$. For each y near x either $g(y) = g(x)$, so $f \circ g(y) - f \circ g(x) = 0$, or (3.5) holds. In either case, $[f \circ g(y) - f \circ g(x)](y - x)^{-1}$ is close to zero for y near x. $\square$

Proposition 3.9. (Change of variables in integration). *Suppose* $g \colon [a, b] \to \mathbb{R}$ *is continuous, and is differentiable at each point of* (a, b). *Suppose* $f \colon g([a, b]) \to \mathbb{C}$ *is continuous. Then*

$$\int_{g(a)}^{g(b)} f = \int_a^b (f \circ g)g'.$$

Proof. Define $F \colon g([a, b]) \to \mathbb{C}$ and $G \colon [a, b] \to \mathbb{C}$ by

$$F(y) = \int_{g(a)}^y f, \qquad G(x) = \int_a^x (f \circ g)g'.$$

We want to prove that $F(g(b)) = G(b)$; we shall in fact show that $F \circ g = G$ on $[a, b]$. Since $F \circ g(a) = G(a) = 0$, it suffices to prove that the derivatives are the same. But

$$(F \circ g)' = (F' \circ g)g' = (f \circ g)g' = G'. \qquad \square$$

A function $f: (a, b) \to \mathbb{C}$ is said to be *differentiable* if it is differentiable at each point of (a, b). If f is differentiable, then the derivative f' is itself a function from (a, b) to $\mathbb{C}$ which may (or may not) have a derivative $f''(x) = (f')'(x)$ at $x \in (a, b)$. This is called the *second derivative* of f at x and denoted also by

$$D^2 f(x), \qquad \frac{d^2 f}{dx^2}(x).$$

Higher derivatives are defined similarly, by induction:

$$f^{(0)}(x) = f(x), \qquad f^{(1)}(x) = f'(x),$$
$$f^{(k+1)}(x) = (f^{(k)})'(x), \qquad k = 0, 1, 2, \dots.$$

The function $f: (a, b) \to \mathbb{C}$ is said to be of *class C^k*, or *k-times continuously differentiable*, if each of the derivatives $f, f', \dots, f^{(k)}$ is a continuous function on (a, b). The function is said to be a *class C^∞*, or *infinitely differentiable*, if $f^{(k)}$ is continuous on (a, b) for every integer $k \geq 0$.

Exercises

1. Show that any polynomial is infinitely differentiable.

2. Show that the Mean Value Theorem is not true for complex-valued functions, in general, by finding a differentiable function f such that $f(0) = 0 = f(1)$ but $f'(x) \neq 0$ for $0 < x < 1$.

3. State and prove a theorem analogous to Theorem 3.7 when $f'(x) < 0$, all $x \in (a, b)$.

4. Suppose f, g are of class C^k and $c \in \mathbb{C}$. Show that $f + g$, cf, and fg are of class C^k.

5. Suppose p is a polynomial with real coefficients. Show that between any two distinct real roots of p there is a real root of p'.

6. Show that for any $k = 0, 1, 2, \dots$ there is a function $f: \mathbb{R} \to \mathbb{R}$ which is of class C^k, such that $f(x) = 0$ if $x \leq 0$, $f(x) > 0$ if $x > 0$. Is there a function of class C^∞ having this property?

7. Prove the following extension of the mean value theorem: if f and g are continuous real-valued functions on $[a, b]$, and if the derivatives exist at each point of (a, b), then there is $c \in (a, b)$ such that

$$[f(b) - f(a)]g'(c) = [g(b) - g(a)]f'(c).$$

8. Prove *L'Hôpital's rule*: if f and g are as in Exercise 7 and if

$$\lim_{x \to a} f'(x)[g'(x)]^{-1}$$

exists and $f(a) = g(a) = 0$ and g, g' are non-zero on $(a, b]$, then

$$\lim_{x \to a} f(x)[g(x)]^{-1}$$

exists and the two limits are equal.

§4. Sequences and series of functions

Suppose that S is any set and that $f: S \to \mathbb{C}$ is a bounded function. Let

$$|f| = \sup \{|f(x)| \mid x \in S\}.$$

A sequence of bounded functions $(f_n)_{n=1}^{\infty}$ from S to $\mathbb{C}$ is said to *converge uniformly* to f if

$$\lim_{n \to \infty} |f_n - f| = 0,$$

This sequence $(f_n)_{n=1}^{\infty}$ is said to be a *uniform Cauchy sequence* if for each $\varepsilon > 0$ there is an integer N so that

(4.1) $|f_n - f_m| < \varepsilon$ if $n, m \geq N$.

It is not difficult to show that if the sequence converges uniformly to a function f, then it is a uniform Cauchy sequence. The converse is also true.

Theorem 4.1. *Suppose* $(f_n)_{n=1}^{\infty}$ *is a sequence of bounded functions from a set S to $\mathbb{C}$ which is a uniform Cauchy sequence. Then there is a unique bounded function $f: S \to \mathbb{C}$ such that* $(f_n)_{n=1}^{\infty}$ *converges uniformly to f. If S is a metric space and each f_n is continuous, then f is continuous.*

Proof.   For each $x \in S$, we have

$$|f_n(x) - f_m(x)| \leq |f_n - f_m|.$$

Therefore $(f_n(x))_{n=1}^{\infty}$ is a Cauchy sequence in $\mathbb{C}$. Denote the limit by $f(x)$. We want to show that $(f_n)_{n=1}^{\infty}$ converges uniformly to the function f defined in this way. Given $\varepsilon > 0$, take N so large that (4.1) holds. Then for a fixed $m \geq N$,

$$|f_m(x) - f(x)| = |f_m(x) - \lim_{n \to \infty} f_n(x)|$$

$$= \lim_{n \to \infty} |f_m(x) - f_n(x)| \leq \varepsilon.$$

Thus $|f_m - f| \leq \varepsilon$ if $m \geq N$, and $(f_n)_{n=1}^{\infty}$ converges uniformly to f. If the sequence also converged uniformly to g, then for each $x \in S$,

$$|f_n(x) - g(x)| \leq |f_n - g|$$

so $f_n(x) \to g(x)$ as $n \to \infty$, and $g = f$. The function f is bounded, since

$$|f(x)| = |f(x) - f_m(x) + f_m(x)| \le \varepsilon + |f_m(x)| \le \varepsilon + |f_m|,$$

so $|f| \le \varepsilon + |f_m|$, if $m \ge N$.

Finally, suppose each f_n is continuous on the metric space S. Suppose $x \in S$ and $\varepsilon > 0$. Choose N as above. Choose $\delta > 0$ so small that

$$|f_N(y) - f_N(x)| < \varepsilon \qquad \text{if } d(y, x) < \delta.$$

Then

$$\begin{aligned}
|f(y) - f(x)| &\le |f(y) - f_N(y)| + |f_N(y) - f_N(x)| + |f_N(x) - f(x)| \\
&\le |f - f_N| + |f_N(y) - f_N(x)| + |f - f_N| \\
&< 3\varepsilon \qquad \text{if } d(y, x) < \delta.
\end{aligned}$$

Thus f is continuous. ☐

The usefulness of the notion of uniform convergence is indicated by the next theorem and the example following.

Theorem 4.2. *Suppose $(f_n)_{n=1}^{\infty}$ is a sequence of continuous complex-valued functions on the interval $[a, b]$, and suppose it converges uniformly to f. Then*

$$\int_a^b f = \lim_{n \to \infty} \int_a^b f_n.$$

Proof. By (2.2),

$$\left| \int_a^b f_n - \int_a^b f \right| = \left| \int_a^b (f_n - f) \right| \le |f_n - f| \cdot |b - a|.$$

As $n \to \infty$, this $\to 0$. ☐

Example

For each positive integer n, let $f_n: [0, 1] \to \mathbb{R}$ be the function whose graph consists of the line segments joining the pairs of points $(0, 0)$, $((2n)^{-1}, 2n)$; $((2n)^{-1}, 2n)$, $(n^{-1}, 0)$; $(n^{-1}, 0)$, $(1, 0)$. Then f_n is continuous, $f_n(x) \to 0$ as $n \to \infty$ for *each* $x \in [0, 1]$, but $\int_0^1 f_n = 1$, all n.

Here we are interested particularly in sequences of functions which are partial sums of power series. Associated with the sequence $(a_n)_{n=0}^{\infty}$ in $\mathbb{C}$ and the point $z_0 \in \mathbb{C}$ is the series

$$(4.2) \qquad \sum_{n=0}^{\infty} a_n(z - z_0)^n, \qquad z \in \mathbb{C}.$$

Recall from §3 of Chapter 1 that there is a number R, $0 \le R \le \infty$, such that (4.2) converges when $|z - z_0| < R$ and diverges when $|z - z_0| > R$; R is called the *radius of convergence* of (3.2). The partial sums

$$(4.3) \qquad f_n(z) = \sum_{m=0}^{n} a_m(z - z_0)^m$$

are continuous functions on $\mathbb{C}$ which converge at each point z with $|z - z_0| < R$ to the function

$$(4.4) \qquad f(z) = \sum_{m=0}^{\infty} c_m(z - z_0)^m, \qquad |z - z_0| < R.$$

Theorem 4.3. *Let R be the radius of convergence of the power series* (4.2). *Then the function f defined by* (4.4) *is a continuous function. Moreover, the functions f_n defined by* (4.3) *converge to f uniformly on each disc*

$$D_r = \{z \mid |z - z_0| < r\}, \qquad 0 < r < R.$$

Proof. We prove uniform convergence first. Given $0 < r < R$, choose s with $r < s < R$. Take w with $|w - z_0| = s$. By assumption, $\sum a_n(w - z_0)^n$ converges. Therefore the terms of this series $\to 0$. It follows that there is a constant M so that

$$|a_n(w - z_0)^n| \le M, \qquad n = 0, 1, \ldots.$$

Since $|w - z_0| = s$, this means

$$(4.5) \qquad |a_n| \le Ms^{-n}, \qquad n = 0, 1, \ldots,$$

Now suppose $z \in D_r$ and $m < n$. Then

$$|f_n(z) - f_m(z)| = \left| \sum_{m+1}^{n} a_j(z - z_0)^j \right|$$

$$\le \sum_{m=1}^{n} |a_j| \, |z - z_0|^j$$

$$\le \sum_{m+1}^{n} Ms^{-j} r^j$$

$$= M \sum_{n+1}^{n} \delta^j = M \frac{\delta^{m+1} - \delta^{n+1}}{1 - \delta} \le \frac{M\delta^{m+1}}{1 - \delta},$$

where $\delta = r/s < 1$. As $m \to \infty$ the final expression on the right $\to 0$, so $(f_n)_{n=0}^{\infty}$ is a uniform Cauchy sequence on D_r. It follows that it converges to f uniformly on D_r and that f is continuous on D_r. Since this is true for each $r < R$, f is continuous. $\square$

In particular, suppose $x_0 \in \mathbb{R}$. The power series

$$(4.6) \qquad \sum_{n=0}^{\infty} a_n(x - x_0)^n$$

defines a continuous function in the open interval $(x_0 - R, x_0 + R)$. Is this function differentiable?

Theorem 4.4. *Suppose the power series* (4.6) *has radius of convergence R. Then the function f defined by this series is differentiable, and*

$$(4.7) \qquad f'(x) = \sum_{n=1}^{\infty} na_n(x - x_0)^{n-1}, \qquad |x - x_0| < R.$$

Proof. To simplify notation we shall assume $x_0 = 0$. We claim first that the two series

(4.8)
$$\sum_{n=1}^{\infty} na_n x^{n-1}, \qquad \sum_{n=2}^{\infty} n^2 a_n x^{n-2}$$

converge uniformly for $|x| \leq r < R$. Take $r < s < R$. Then (4.5) holds. It follows that

$$\sum_{m=1}^{n} |ma_m x^{m-1}| \leq M \sum_{m=1}^{n} ms^{-m} r^{m-1}$$
$$= Mr^{-1} \sum_{m=1}^{n} m\delta^m,$$

$\delta = r/s < 1$. Take $\varepsilon > 0$ so small that $(1 + \varepsilon)\delta < 1$. By Exercise 4 of Chapter 1, §3, $m \leq (1 + \varepsilon)^m$ for all large m. Therefore there is a constant M' so that

$$m \leq M'(1 + \varepsilon)^m, \qquad m = 1, 2, \ldots .$$

Then

$$\sum_{m=1}^{n} |ma_m x^{m-1}| \leq r^{-1}MM' \sum_{m=1}^{n} (1 + \varepsilon)^m \delta^m.$$

This last series converges, so the first series in (4.8) converges uniformly for $|x| \leq r$. Similarly, $m^2 \leq (1 + \varepsilon)^m$ for large m, and the second series in (4.8) converges uniformly for $|x| \leq r$.

Let g be the function defined by the first series in (4.8). Recall that we are taking x_0 to be 0. We want to show that

(4.9) $[f(y) - f(x)](y - x)^{-1} - g(x) \to 0 \quad$ as $\quad y \to x.$

Assume $|x| < r, |y| < r$. Then the expression in (4.9) is

(4.10)
$$\sum_{n=2}^{\infty} a_n[y^n - x^n - nx^{n-1}(y - x)](y - x)^{-1}.$$

Now

$$y^n - x^n = (y - x)g_n(x, y)$$

where

$$g_n(x, y) = y^{n-1} + xy^{n-2} + \cdots + x^{n-2}y + x^{n-1}.$$

Thus

$$|g_n(x, y)| \leq nr^{n-1} \qquad \text{if } |x| \leq r, \quad |y| \leq r.$$

Then

$$y^n - x^n - nx^{n-1}(y - x)$$
$$= (y - x)[y^{n-1} + xy^{n-2} + \cdots + x^{n-1} - nx^{n-1}]$$
$$= (y - x)[(y^{n-1} - x^{n-1}) + (y^{n-2} - x^{n-2})x + \cdots + (y - x)x^{n-2}$$
$$\qquad\qquad\qquad\qquad\qquad\qquad + x^{n-1} - x^{n-1}]$$
$$= (y - x)^2[g_{n-1}(x, y) + g_{n-2}(x, y)x + \cdots + g_1(x, y)x^{n-2}],$$

so

$$|y^n - x^n - nx^{n-1}(y - x)| \le |y - x|^2 \cdot n^2 r^{n-2}.$$

It follows that for $|x| \le r, |y| \le r$ we have

$$|[f(y) - f(x)](y - x)^{-1} - g(x)| \le \sum_{n=2}^{\infty} |a_n| |y - x|^2 n^2 r^{n-2} |y - x|^{-1}$$

$$= |y - x| \sum_{n=2}^{\infty} n^2 |a_n| r^{n-2} = K|y - x|,$$

K constant. Thus $f'(x) = g(x)$. ☐

Corollary 4.5. *The function f in Theorem 4.4 is infinitely differentiable, and*

$$(4.11) \qquad f^{(k)}(x) = \sum_{n=k}^{\infty} n(n - 1)(n - 2) \cdots (n - k + 1)a_n(x - x_0)^{n-k},$$

$|x - x_0| < R$.

Proof. This follows from Theorem 4.4 by induction on k. ☐

In particular, if we take $x = x_0$ in (4.11) then all terms of the series except the first are zero and (4.11) becomes

$$(4.12) \qquad\qquad a_k = (k!)^{-1} f^{(k)}(x_0).$$

This means that the *coefficients of the power series* (4.6) *are determined uniquely by the function f* (provided the radius of convergence is positive).

Exercises

1. Find the function defined for $|x| < 1$ by $f(x) = \sum_{n=1}^{\infty} x^n/n$. (Hint: $f(x) = \int_0^x f'$.)

2. Show that if f is defined by (4.6), then

$$\int_{x_0}^x f = \sum_{n=0}^{\infty} (n + 1)^{-1} a_n(x - x_0)^{n+1}.$$

3. Find the function defined for $|x| < 1$ by $f(x) = \sum_{n=1}^{\infty} nx^{n-1}$.

4. Suppose there is a sequence $(x_n)_{n=1}^{\infty}$ such that $|x_{n+1} - x_0| < |x_n - x_0|$, $x_n \to x_0$, and $f(x_n) = 0$ for each n, where f is defined by (4.6). Show that $f(x) = 0$ for *all* x. (Hint: show that $a_0, a_1, a_2, \ldots$ are each zero.)

§5. Differential equations and the exponential function

Rather than define the general notion of a "differential equation" here, we shall consider some particular examples. We begin with the problem of finding a continuously differentiable function $E: \mathbb{R} \to \mathbb{C}$ such that

$$(5.1) \qquad\qquad E(0) = 1, \qquad E'(x) = E(x), \qquad x \in \mathbb{R}.$$

Suppose there were such a function E, and suppose it could be defined by a power series

$$E(x) = \sum_{n=0}^{\infty} a_n x^n.$$

Then (5.1) and Theorem 4.4 imply

$$\sum_{n=1}^{\infty} n a_n x^{n-1} = \sum_{n=0}^{\infty} a_n x^n,$$

or

$$\sum_{n=0}^{\infty} (n+1) a_{n+1} x^n = \sum_{n=0}^{\infty} a_n x^n.$$

Since the coefficients are uniquely determined, this implies

$$a_{n+1} = a_n/(n+1), \qquad n = 0, 1, 2, \ldots.$$

But

$$a_0 = E(0) = 1,$$

so inductively

$$a_n = (n!)^{-1} = [n(n-1)(n-2)\cdots 1]^{-1}.$$

We have shown that if there is a solution of (5.1) defined by a power series, then it is given by

(5.2) $$E(x) = \sum_{n=0}^{\infty} (n!)^{-1} x^n.$$

The ratio test shows that (5.2) converges for all real or complex x, and application of Theorem 4.4 shows that E is indeed a solution of (5.1). We shall see that it is the *only* solution.

Theorem 5.1. *For each $a, c \in \mathbb{C}$ there is a unique continuously differentiable function $f: \mathbb{R} \to \mathbb{C}$ such that*

(5.3) $$f(0) = c, \qquad f'(x) = af(x), \qquad x \in \mathbb{R}.$$

This function is

(5.4) $$f(x) = cE(ax) = c \sum_{n=0}^{\infty} (n!)^{-1} a^n x^n.$$

Proof. The function given by (5.4) can be found by the argument used to find E, and Theorem 4.4 shows that it is a solution of (5.3). To show uniqueness, suppose f is any solution of (5.3), and let $g(x) = E(-ax)$, so that

$$g(0) = 1, \qquad g'(x) = -ag(x).$$

Then fg is differentiable and

$$(fg)' = f'g + fg' = afg - afg = 0.$$

Therefore fg is constant and this constant value is $f(0)g(0) = c$. If $c \neq 0$, this implies fg is never zero, so g is never zero. Then for any c,

$$f(x) = c/g(x), \quad \text{all } x.$$

Thus f is unique. □

This can be extended to more complicated problems.

Theorem 5.2. *For each $a, c \in \mathbb{C}$ and each continuous function $h: \mathbb{R} \to \mathbb{C}$ there is a unique continuously differentiable function $f: \mathbb{R} \to \mathbb{C}$ such that*

$$(5.5) \qquad f(0) = c, \qquad f'(x) = af(x) + h(x), \qquad x \in \mathbb{R}.$$

Proof. Let $f_0(x) = E(ax)$, $g(x) = E(-ax)$. As in the preceding proof, $f_0 g$ is constant, $\equiv 1$. Therefore neither function vanishes. Any solution f of (5.5) can be written as

$$f = f_1 f_0, \quad \text{where} \quad f_1 = gf.$$

Then

$$f' = f_1' f_0 + f_1 f_0' = f_1' f_0 + af,$$

so (5.5) holds if and only if

$$f_1(0) = c, \qquad f_1'(x)f_0(x) = h(x).$$

These conditions are equivalent to

$$f_1(x) = c + \int_0^x gh.$$

Thus the unique solution of (5.5) is given by

$$(5.6) \qquad f(x) = cf_0(x) + f_0(x) \int_0^x gh$$

$$= cE(ax) + E(ax) \int_0^x E(-at)h(t)\, dt. \qquad □$$

Now we consider equations involving the second derivative as well.

Theorem 5.3. *For any b, c, d_0, $d_1 \in \mathbb{C}$ and any continuous function $h: \mathbb{R} \to \mathbb{C}$ there is a unique $f: \mathbb{R} \to \mathbb{C}$ of class C^2 which satisfies*

$$(5.7) \qquad f(0) = d_0, \qquad f'(0) = d_1,$$

$$(5.8) \qquad f''(x) + bf'(x) + cf(x) = h(x), \qquad x \in \mathbb{R}.$$

Proof. To motivate the proof, we introduce two operations on functions of class C^1 from $\mathbb{R}$ to $\mathbb{C}$: given such a function g, let

$$Dg = g', \qquad Ig = g.$$

If g is of class C^2, let $D^2 g = g''$. Then (5.8) can be written

$$D^2 f + bDf + cIf = h.$$

This suggests the polynomial $z^2 + bz + c$. We know there are roots a_1, a_2 of this polynomial such that

$$z^2 + bz + c = (z - a_1)(z - a_2), \quad \text{all } z \in \mathbb{C}.$$

Thus

$$b = -a_1 - a_2, \quad c = a_1 a_2.$$

From the properties of differentiation it follows that

$$(D - a_1 I)[(D - a_2 I)f] = D^2 f - (a_1 + a_2)Df + a_2 a_1 f$$
$$= f'' + bf' + cf.$$

Let

$$g = f' - a_2 f = (D - a_2 I)f.$$

We have shown that f is a solution of (5.8) if and only if

$$(D - a_1 I)g = g' - a_1 g = h.$$

If also (5.7) holds, then $g(0) = f'(0) - a_2 f(0) = d_1 - a_2 d_0$. Thus f is a solution of (5.7), (5.8) if and only if

(5.9) $f(0) = d_0, \quad f' - a_2 f = g,$

where

(5.10) $g(0) = d_1 - a_2 d_0, \quad g' - a_1 g = h.$

But (5.10) has a unique solution g, and once g has been found then (5.9) has a unique solution. It follows that (5.7), (5.8) has a unique solution. ☐

Now we return to the function E,

$$E(z) = \sum_{n=0}^{\infty} (n!)^{-1} z^n, \quad z \in \mathbb{C}.$$

Define the real number e by

(5.11) $e = E(1) = \sum_{n=0}^{\infty} (n!)^{-1}.$

Theorem 5.4. *The function E is a function from $\mathbb{R}$ to $\mathbb{R}$ of class C^∞. Moreover,*

(a) $E(x) > 0, x \in \mathbb{R}$,
(b) *for each $y > 0$, there is a unique $x \in \mathbb{R}$ such that $E(x) = y$,*
(c) $E(x + y) = E(x)E(y)$, *all $x, y \in \mathbb{R}$,*
(d) *for any rational r, $E(r) = e^r$.*

Proof. Since E is defined by a power series, it is of class C^∞. It is clearly real for x real, and positive when $x \geq 0$. As above, $E(x)E(-x) = 1$, all x, so also $E(x) > 0$ when $x < 0$.

To prove (b), we wish to apply Theorem 1. Taking the first two terms in the series shows (since $y > 0$) that $E(y) > 1 + y > y$. Also, $E(y^{-1}) > y^{-1}$, so

$$E(-y^{-1}) = E(y^{-1})^{-1} < (y^{-1})^{-1} = y.$$

Thus there is $x \in (-y^{-1}, y)$ such that $E(x) = y$. Since $E' = E > 0$, E is strictly increasing and x is unique.

We have proved (c) when $x = -y$. Multiplying by $E(-x)E(-y)$, we want to show

$$E(x + y)E(-x)E(-y) = 1, \quad \text{all } x, y \in \mathbb{R}.$$

Fix y. This equation holds when $x = 0$, and differentiation with respect to x shows that the left side is constant.

Finally, repeated use of (c) shows that

$$E(nx) = E(x)^n, \quad n = 0, \pm 1, \pm 2, \ldots.$$

Thus

$$e = E(1) = E(n/n) = E(1/n)^n,$$
$$e^{1/n} = E(1/n), \quad n = 1, 2, 3, \ldots.$$
$$e^{m/n} = (e^{1/n})^m = E(1/n)^m = E(m/n). \quad \square$$

Because of (d) above and the continuity of E, it is customary to *define* arbitrary complex powers of e by

(5.12) $$e^z = E(z) = \sum_{n=0}^{\infty} (n!)^{-1} z^n.$$

The notation

(5.13) $$e^z = \exp z$$

is also common.

We extend part of Theorem 5.4 to the complex exponential function. (Recall that z^* denotes the complex conjugate of $z \in \mathbb{C}$.)

Theorem 5.5. *For any complex numbers z and w,*

(5.14) $$E(z + w) = E(z)E(w), \quad E(z^*) = E(z)^*.$$

Proof. The second assertion can be proved by examining the partial sums of the series. To prove the first assertion, recall that we showed in the proof of Theorem 5.1 that $E(zx)E(-zx)$ is constant, $x \in \mathbb{R}$. Therefore $E(z)E(-z) = E(0)^2 = 1$. We want to show

$$E(z + w)E(-z)E(-w) = 1, \quad \text{all } z, w \in \mathbb{C}.$$

Let

$$g(x) = E((z + w)x)E(-zx)E(-wx), \quad x \in \mathbb{R}.$$

Differentiation shows g is constant. But $g(0) = 1$. $\square$

The notation (5.12) and the identity (5.14) can be used to consolidate expressions for the solutions of the differential equations above. The unique solution of

$$f(0) = c, \qquad f' = af + h$$

is

(5.15) $$f(x) = ce^{ax} + \int_0^x e^{a(x-t)}h(t)\, dt.$$

The unique solution of

$$f(0) = c_0, \qquad f'(0) = d_1, \qquad f'' + df' + cf = h$$

is given by

(5.16) $$f(x) = d_0 e^{a_2 x} + \int_0^x e^{a_2(x-t)}g(t)\, dt,$$

where

(5.17) $$g(x) = (d_1 - a_2 d_0)e^{a_1 x} + \int_0^x e^{a_1(x-t)}h(t)\, dt,$$

and a_1, a_2 are the roots of $z^2 + bz + c$.

Exercises

1. Find the solution of

$$f(0) = 1, \qquad f'(0) = 0, \qquad f'' - 2f' + f = 0$$

by the procedure in Theorem 5.3, and also by determining the coefficients in the power series expansion of f.

2. Let f, g be the functions such that

$$f(0) = 1, \qquad f'(0) = 0, \qquad f'' + bf' + cf = 0,$$
$$g(0) = 0, \qquad g'(0) = 1, \qquad g'' + bg' + cg = 0.$$

Show that for any constants d_1, d_2 the function $h = d_1 f + d_2 g$ is a solution of

(*) $$h'' + bh' + ch = 0.$$

Show that conversely if h is a solution of this equation then there are unique constants $d_1, d_2 \in \mathbb{C}$ such that $h = d_1 f + d_2 g$. (This shows that the set of solutions of (*) is a two-dimensional complex vector space, and (f, g) is a basis.)

3. Suppose $h(x) = \sum_{n=0}^{\infty} d_n x^n$, the series converging for all x. Show that the solution of

$$f(0) = 0 = f'(0), \qquad f'' + bf' + cf = h$$

is of the form $\sum_{n=0}^{\infty} a_n x^n$, where this series converges for all x. (Hint: determine the coefficients $a_0, a_1, \ldots$ inductively, and prove convergence.)

4. Suppose $z^3 + bz^2 + cz + d = (z - a_1)(z - a_2)(z - a_3)$, all $z \in \mathbb{C}$. Discuss the problem of finding a function f such that

$$f(0) = e_0, \qquad f'(0) = e_1, \qquad f''(0) = e_2, \qquad f''' + df'' + cf' + d = 0.$$

§6. Trigonometric functions and the logarithm

In §5 the exponential function arose naturally from study of the differential equation $f' = f$. In this section we discuss solutions of one of the simplest equations involving the second derivative: $f'' + f = 0$.

Theorem 6.1. *There are unique functions $S, C: \mathbb{R} \to \mathbb{C}$ of class C^2 such that*

(6.1) $S(0) = 0, \qquad S'(0) = 1, \qquad S'' + S = 0,$

(6.2) $C(0) = 1, \qquad C'(0) = 0, \qquad C'' + C = 0.$

Proof. Existence and uniqueness of such functions is a consequence of Theorem 5.3. □

Let us obtain expressions for S and C using the method of Theorem 5.3. The roots of $z^2 + 1$ are $z = \pm i$. Therefore

$$S(x) = \int_0^x e^{-i(x-t)} g(t) \, dt$$

where

$$g(x) = e^{ix}.$$

Thus

$$S(x) = \int_0^x e^{-i(x-t)} e^{it} \, dt = e^{-ix} \int_0^x e^{2it} \, dt$$

$$= e^{-ix} (2i)^{-1} e^{2it} \big|_0^x = (2i)^{-1} e^{-ix}(e^{2ix} - 1),$$

(6.3) $$S(x) = \frac{1}{2i}(e^{ix} - e^{-ix}).$$

A similar calculation gives

(6.4) $$C(x) = \tfrac{1}{2}(e^{ix} + e^{-ix}).$$

Theorem 6.2. *The functions S, C defined by (6.3) and (6.4) are real-valued functions of class C^∞ on $\mathbb{R}$. Moreover,*

(a) $S' = C, C' = -S,$
(b) $S(x)^2 + C(x)^2 = 1,$ *all* $x \in \mathbb{R}$,

(c) *there is a smallest positive number p such that* $C(p) = 0$,

(d) *if p is the number in* (c), *then*

$$S(x + 4p) = S(x), \qquad C(x + 4p) = C(x), \qquad all \ x \in \mathbb{R}.$$

Proof. Since the exponential function $\exp(ax)$ is of class C^∞ as a function of x for each $a \in \mathbb{C}$, S and C are of class C^∞. Since $(\exp(ix))^* = \exp(-ix)$, S and C are real-valued. In fact,

$$C(x) = \mathrm{Re}\,(e^{ix}), \qquad S(x) = \mathrm{Im}\,(e^{ix}),$$

so

$$e^{ix} = C(x) + iS(x).$$

Differentiation of (6.3) and (6.4) shows $S' = C$, $C' = -S$. Differentiation of $S^2 + C^2$ shows that $S(x)^2 + C(x)^2$ is constant; the value for $x = 0$ is 1.

To prove (c), we suppose that $C(x) \neq 0$ for all $x > 0$. Since $C(0) = 1$ and C is continuous, the Intermediate Value Theorem implies $C(x) > 0$, all $x > 0$. Since $S' = C$, S is then strictly increasing for $x \geq 0$. In particular, $S(x) \geq S(1) > 0$, all $x \geq 1$. But then

$$0 < C(x) = C(1) + \int_1^x C'(t)\,dt = C(1) - \int_1^x S(t)\,dt$$

$$\leq C(1) - \int_1^x S(1)\,dt = C(1) - (x - 1)S(1), \qquad x \geq 1.$$

But for large x the last expression is negative, a contradiction. Thus $C(x) = 0$ for some $x > 0$. Let $p = \inf\{x \mid x > 0, C(x) = 0\}$. Then $p \geq 0$. There is a sequence $(x_n)_1^\infty$ such that $C(x_n) = 0, p \leq x_n \leq p + 1/n$. Thus $C(p) = 0$, and p is the smallest positive number at which C vanishes.

To prove (d) we note that

$$1 = S(p)^2 + C(p)^2 = S(p)^2,$$

so $S(p) = \pm 1$. But $S(0) = 0$ and $S' = C$ is positive on $[0, p)$, so $S(p) > 0$, Thus $S(p) = 1$. Consider $S(x + p)$ as a function of x. It satisfies (6.2), and so by uniqueness we must have

$$S(x + p) = C(x), \qquad x \in \mathbb{R}.$$

Similarly, $-C(x + p)$ considered as a function of x satisfies (6.1), so

$$C(x + p) = -S(x), \qquad x \in \mathbb{R}.$$

Then

$$S(x + 4p) = C(x + 3p) = -S(x + 2p) = -C(x + p) = S(x),$$
$$C(x + 4p) = -S(x + 3p) = -C(x + 2p) = S(x + p) = C(x). \qquad \Box$$

We *define* the positive number π by

$$\pi = 2p,$$

p the number in (c), (d) of Theorem 6.3. We *define* the functions sine and cosine for all $z \in \mathbb{C}$ by

$$(6.5) \qquad \sin z = \frac{1}{2i}(e^{iz} - e^{-iz}) = \sum_{n=0}^{\infty} [(2n+1)!]^{-1}(-1)^n z^{2n+1},$$

$$(6.6) \qquad \cos z = \frac{1}{2}(e^{iz} + e^{-iz}) = \sum_{n=0}^{\infty} [2n!]^{-1}(-1)^n z^{2n}.$$

Note that because of the way we have defined π and the sine and cosine functions, it is necessary to *prove* that they have the usual geometric significance.

Theorem 6.3. *Let $\gamma \colon [0, 2\pi] \to \mathbb{R}^2$ be defined by*

$$\gamma(t) = (\cos t, \sin t).$$

Then γ is a 1-1 mapping of $[0, 2\pi]$ onto the unit circle about the origin in $\mathbb{R}^2$. The length of the arc of this circle from $\gamma(0)$ to $\gamma(t)$ is t. In particular, the length of the unit circle is 2π.

Proof. We know from Theorem 6.2 that $(\cos t)^2 + (\sin t)^2 = 1$, so $(\cos t, \sin t)$ lies on the unit circle. The discussion in the proof of Theorem 6.2 shows that on the interval $[0, \frac{1}{2}\pi]$, $\cos t$ decreases strictly from 1 to 0 while $\sin t$ increases strictly from 0 to 1. Therefore, γ maps $[0, \frac{1}{2}\pi]$ *into* the portion of the circle lying in the quadrant $x \geq 0, y \geq 0$ in a 1-1 manner. Furthermore, suppose $0 \leq x \leq 1, 0 \leq y \leq 1$, and $x^2 + y^2 = 1$. By the Intermediate Value Theorem and the continuity of cosine, there is a unique $t \in [0, \frac{1}{2}\pi]$ such that $\cos t = x$. Then $\sin t \geq 0$, $(\sin t)^2 = 1 - x^2 = y^2$, and $y \geq 0$, so $\sin t = y$. Thus γ maps $[0, \frac{1}{2}\pi]$ *onto* the portion of the circle in question.

Since $\cos(t + \frac{1}{2}\pi) = -\sin t$ and $\sin(t + \frac{1}{2}\pi) = \cos t$, the cosine decreases from 0 to -1 and the sine decreases from 1 to 0 on $[\frac{1}{2}\pi, \pi]$. As above, we find that γ maps this interval 1-1 and onto the portion of the circle in the quadrant $x \leq 0, y \geq 0$. Continuing in this way we see that γ does indeed map $[0, 2\pi]$ 1-1 onto the unit circle.

The length of the curve γ from $\gamma(0)$ to $\gamma(t)$ is usually *defined* to be the limit, if it exists, of the lengths of polygonal approximations. Specifically, suppose

$$0 = t_0 < t_1 < t_2 < \cdots < t_n = t.$$

The sum of the lengths of the line segments joining the points $\gamma(t_{i-1})$ and $\gamma(t_i)$, $i = 1, 2, \ldots, n$ is

$$(6.7) \qquad \sum_{i=1}^{n} [(\cos t_i - \cos t_{i-1})^2 + (\sin t_i - \sin t_{i-1})^2]^{1/2}.$$

By the Mean Value Theorem, there are t_i' and t_i'' between t_{i-1} and t_i such that

$$\cos t_i - \cos t_{i-1} = -\sin t_i'(t_i - t_{i-1}),$$
$$\sin t_i - \sin t_{i-1} = \cos t_i''(t_i - t_{i-1}).$$

Therefore, the sum (6.7) is

$$\sum_{i=1}^{n} [(\sin t_i')^2 + (\cos t_i'')^2](t_i - t_{i-1}).$$

Since sine and cosine are continuous, hence uniformly continuous on $[0, t]$, and since $(\sin t)^2 + (\cos t)^2 = 1$, it is not hard to show that as the maximum length $|t_i - t_{i-1}| \to 0$, (6.7) approaches t. $\square$

This theorem shows that sine, cosine, and π as defined above do indeed have the usual interpretation. Next we consider them as functions from $\mathbb{C}$ to $\mathbb{C}$.

Theorem 6.4. *The sine, cosine, and exponential functions have the following properties:*

(a) $\exp(iz) = \cos z + i \sin z$, *all* $z \in \mathbb{C}$,
(b) $\sin(z + 2\pi) = \sin z$, $\cos(z + 2\pi) = \cos z$, $\exp(z + 2\pi i) = \exp z$, *all* $z \in \mathbb{C}$,
(c) *if* $w \in \mathbb{C}$ *and* $w \neq 0$, *there is a* $z \in \mathbb{C}$ *such that* $w = \exp(z)$. *If also* $w = \exp(z')$, *then there is an integer n such that* $z' = z + 2n\pi i$.

Proof. The identity (a) follows from solving (6.5) and (6.6) for $\exp(iz)$. By Theorem 2.2 and the definition of π,

$$\exp(2\pi i) = \cos 2\pi + i \sin 2\pi = 1.$$

Then since $\exp(z + w) = \exp z \exp w$ we get

$$\exp(z + 2\pi i) = \exp z.$$

This identity and (6.5), (6.6) imply the rest of (b).

Suppose $w \in \mathbb{C}$, $w \neq 0$. Let $r = |w|$. If x, y are real,

$$|\exp(iy)|^2 = |\cos y + i \sin y|^2 = (\cos y)^2 + (\sin y)^2 = 1.$$

Therefore

$$|\exp(x + iy)|^2 = |\exp x \exp(iy)| = |\exp x| = \exp x.$$

To have $\exp(x + iy) = w$, then, we must have $\exp x = r$. By Theorem 5.4 there is a unique such $x \in \mathbb{R}$. We also want $\exp(iy) = r^{-1}w = a + bi$. Since $|r^{-1}w| = 1$, $a^2 + b^2 = 1$. By Theorem 6.3 there is a unique $y \in [0, 2\pi)$ such that $\cos y = a$, $\sin y = b$. Then $\exp(iy) = \cos y + i \sin y = a + ib$. We have shown that there are $x, y \in \mathbb{R}$ such that if $z = x + iy$,

$$\exp z = \exp x \exp(iy) = r \cdot r^{-1} w = w.$$

Suppose $z' = x' + iy'$, x', y' real, and $\exp z' = w$. Then

$$r = |w| = |\exp z'| = |\exp x'| = \exp x',$$

$x' = x$. There is an integer n such that

$$2n\pi \leq y' - y < 2n\pi + 2\pi,$$

or

$$y' = y + 2n\pi + h, \qquad h \in [0, 2\pi).$$

Since $\exp z' = \exp z$, we have

$$\exp(iy) = \exp(iy') = \exp(i(y + 2n\pi + h)) = \exp(iy + ih),$$

so

$$1 = \exp(-iy)\exp(iy + ih) = \exp(ih).$$

Since $h \in [0, 2\pi)$, this implies $h = 0$. Thus $y' = y + 2n\pi$. ☐

The trigonometric functions tangent, secant, etc., are defined for complex values by

$$\tan z = \sin z/\cos z, \qquad z \in \mathbb{C}, \qquad \cos z \neq 0,$$

etc.

If $w, z \in \mathbb{C}$ and $w = \exp z$, then z is said to be a *logarithm* of w,

$$z = \log w.$$

Theorem 6.4 shows that any $w \neq 0$ has a logarithm; in fact it has infinitely many, whose imaginary parts differ by integral multiples of 2π. Thus $\log w$ is not a function of w, in the usual sense. It can be considered a function by restricting the range of values of the imaginary part. For example, if $w \neq 0$ the z such that $\exp z = w$, $\mathrm{Im}\, z \in [a, a + 2\pi)$ is unique, for any given choice of $a \in \mathbb{R}$.

If $x > 0$, it is customary to take for $\log x$ the unique *real* y such that $\exp y = x$. Thus as a function from $(0, \infty)$ to $\mathbb{R}$, the logarithm is the inverse of the exponential function. Theorem 3.7 shows that it is differentiable, with

$$\frac{d}{dx}(\log x) = \left(\frac{d}{dy}e^y\Big|_{y=\log x}\right)^{-1} = e^{-\log x} = x^{-1}.$$

Thus

(6.8) $$\log x = \int_1^x t^{-1}\,dt, \qquad x > 0.$$

Exercises

1. Prove the identities

$$\sin(z + w) = \sin z \cos w + \cos z \sin w,$$
$$\cos(z + w) = \cos z \cos w - \sin z \sin w$$

for all complex z, w. (Hint: use (6.5) and (6.6).)

2. Show that $\tan x$ is a strictly increasing function from $(-\tfrac{1}{2}\pi, \tfrac{1}{2}\pi)$ onto $\mathbb{R}$. Show that the inverse function $\tan^{-1} x$ satisfies

$$\frac{d}{dx}(\tan^{-1} x) = (1 + x^2)^{-1}$$

3. Show that $\int_{-\infty}^{\infty} (1 + x^2)^{-1}\, dx = \pi$.

4. Show that

$$\log(1 + x) = \int_0^x (1 + t)^{-1}\, dt, \qquad -1 < x < \infty.$$

5. Show that

$$\log(1 + x) = \sum_{n=1}^{\infty} \frac{(-1)^{n-1}}{n} x^n, \qquad -1 < x < 1.$$

(Hint: use Exercise 4.)

§7. Functions of two variables

Suppose A is an open subset of $\mathbb{R}^2$, i.e., for each $(x_0, y_0) \in A$ there is an open disc with center (x_0, y_0) contained in A:

$$A \supset \{(x, y) \mid (x - x_0)^2 + (y - y_0)^2 < r^2\}, \qquad \text{some } r > 0.$$

In particular, A contains (x, y_0) for each x in the open interval $x_0 - r < x < x_0 + r$, and A contains (x_0, y) for each y in the open interval $y_0 - r < y < y_0 + r$.

Suppose $f: A \to \mathbb{C}$. It makes sense to ask whether $f(x, y_0)$ is differentiable as a function of x at x_0. If so, we denote the derivative by

$$D_1 f(x_0, y_0) = \lim_{x \to x_0} [f(x, y_0) - f(x_0, y_0)](x - x_0)^{-1}.$$

Other common notations are

$$\frac{\partial f}{\partial x}(x_0, y_0), \qquad \frac{\partial f}{\partial x}\bigg|_{(x_0, y_0)}, \qquad f_x(x_0, y_0), \qquad D_x f(x_0, y_0).$$

Similarly, if $f(x_0, y)$ is differentiable at y_0 as a function of y we set

$$D_2 f(x_0, y_0) = \lim_{y \to y_0} [f(x_0, y) - f(x_0, y_0)](y - y_0)^{-1}.$$

The derivatives $D_1 f$, $D_2 f$ are called the *first order partial derivatives* of f. The *second order* partial derivatives are the first order derivatives of the first order derivatives:

$$D_1^2 f = D_1(D_1 f), \qquad D_2^2 f = D_2(D_2 f),$$
$$D_1 D_2 f = D_1(D_2 f), \qquad D_2 D_1 f = D_2(D_1 f).$$

Other notations are

$$\frac{\partial^2 f}{\partial x^2}, \quad \frac{\partial^2 f}{\partial y^2}, \quad \frac{\partial^2 f}{\partial x\,\partial y}, \quad \frac{\partial^2 f}{\partial y\,\partial x}, \quad \text{etc.}$$

Higher order partial derivatives are defined similarly. An $(n + 1)$-order partial derivative of f is $D_1 g$ or $D_2 g$, where g is an n-order partial derivative of f. The function $f: A \to \mathbb{C}$ is said to be *n-times continuously differentiable*, or *of class C^n* if all the partial derivatives of f of order $\le n$ exist and are continuous at each point of A. If this is true for every integer n, then f is said to be *infinitely differentiable*, or *of class C^∞*.

Theorem 7.1. (Equality of mixed partial derivatives). *If $f: A \to \mathbb{C}$ is of class C^2, then $D_1 D_2 f = D_2 D_1 f$.*

Proof. Suppose $(a, b) \in A$. Choose $r > 0$ so small that A contains the closed square with center (a, b), edges parallel to the coordinate axes, and sides of length $2r$. Thus $(x, y) \in A$ if

$$|x - a| \le r \quad \text{and} \quad |y - b| \le r.$$

In this square we apply the fundamental theorem of calculus to f as a function of x with y fixed, and conclude

$$f(x, y) = \int_a^x D_1 f(s, y)\, ds + f(a, y).$$

Let $g(y) = f(a, y)$. We claim that

$$(7.1) \qquad D_2 f(x, y) = \int_a^x D_2 D_1 f(s, y)\, ds + g'(y).$$

If so, then differentiation with respect to x shows $D_1 D_2 f = D_2 D_1 f$. To prove (7.1) we consider

$$\varepsilon^{-1}(f(x, y + \varepsilon) - f(x, y)) - \int_a^x D_2 D_1 f(s, y)\, ds - g'(y)$$

$$= \int_a^x [\varepsilon^{-1}(D_1 f(s, y + \varepsilon) - D_1 f(s, y)) - D_2 D_1 f(s, y)]\, ds$$

$$+ [\varepsilon^{-1}(g(y + \varepsilon) - g(y)) - g'(y)].$$

The second term in brackets $\to 0$ as $\varepsilon \to 0$. If f is real-valued, we may apply the Mean Value Theorem to the first term and conclude that for each s and y and for each small ε, there is a point $y' = y'(s, y, \varepsilon)$ between y and $y + \varepsilon$ such that

$$(7.2) \quad \varepsilon^{-1}(D_1 f(s, y + \varepsilon) - D_1 f(s, y)) - D_2 D_1 f(s, y)$$
$$= D_2 D_1 f(s, y') - D_2 D_1 f(s, y).$$

Now $|y' - y| < \varepsilon$, so $|(s, y') - (s, y)| < \varepsilon$. Since $D_2 D_1 f$ is uniformly continuous on the square $|x - a| \le r, |y - b| \le r$, it follows that the maximum value of (7.2) converges to zero as $\varepsilon \to 0$. This implies convergence to zero of the integral of (7.1) with respect to s, proving (7.1) when f is real. In the general case, we look at the real and imaginary parts of f separately. □

Remarks. In the course of proving Theorem 7.1 we have, in effect, proved the following. If f is a complex-valued function of class C^1 defined on a rectangle $|x - a| < r_1$, $|y - b| < r_2$, then the derivative with respect to y of

$$\int_a^x f(s, y) \, ds$$

is

$$\int_a^x D_2 f(s, y) \, ds.$$

Similarly, the derivative with respect to x of

$$\int_b^y f(x, t) \, dt$$

is

$$\int_b^y D_1 f(x, t) \, dt.$$

We need one more result of this sort: if $a = b$, $r_1 = r_2$, then

$$F(y) = \int_a^y f(s, y) \, ds$$

is defined for $|y - a| < r_1$. The derivative is

(7.3)
$$\int_a^y D_2 f(s, y) \, ds + f(y, y).$$

In fact

$$F(y + \varepsilon) - F(y) = \int_a^y [f(s, y + \varepsilon) - f(s, y)] \, ds + \int_y^{y+\varepsilon} f(s, y + \varepsilon) \, dy.$$

Divide by ε and let $\varepsilon \to 0$. By the argument above, the first integral converges to

$$\int_a^y D_2 f(s, y) \, ds.$$

In the second integral, we are integrating a function whose values are very close to $f(y, y)$, over an interval of length ε. Then, dividing by ε, we see that the limit is $f(y, y)$.

We need two results on change of order of integration.

Theorem 7.2. *Suppose f is a continuous complex valued function on the rectangle*

$$A = \{(x, y) \mid a \le x \le b, c \le y \le d\}.$$

Then the functions

$$g(x) = \int_c^d f(x, t)\, dt, \qquad h(y) = \int_a^b f(s, y)\, ds$$

are continuous, and

$$\int_a^b g(x)\, dx = \int_c^d h(y)\, dy.$$

Proof. The preceding remarks show that g and h are not only continuous but differentiable. More generally,

$$\int_b^y f(s, t)\, dt, \qquad \int_a^x f(s, t)\, ds$$

are continuous functions of s and of t respectively. Define

$$F_1(x, y) = \int_a^x \left\{ \int_c^y f(s, t)\, dt \right\} ds,$$

$$F_2(x, y) = \int_c^y \left\{ \int_a^x f(s, t)\, ds \right\} dt.$$

We want to show that $F_1(b, d) = F_2(b, d)$. The remarks preceding this theorem show that

$$D_2 F_1(x, y) = \int_a^x f(s, y)\, ds = D_2 F_2(x, y).$$

Therefore, $F_2 - F_1$ is constant along each vertical line segment in the rectangle A. Similarly, $D_1 F_1 = D_1 F_2$, so $F_2 - F_1$ is constant along each horizontal line segment. Since $F_1(a, c) = F_2(a, c) = 0$, $F_1 \equiv F_2$. $\square$

The next theorem describes the analogous situation for integration over a triangle.

Theorem 7.3. *Suppose f is a continuous complex-valued function defined on the triangle*

$$A = \{(x, y) \mid 0 \le x \le a, 0 \le y \le x\}.$$

Then

$$\int_0^a \left\{ \int_0^x f(x, y)\, dy \right\} dx = \int_0^a \left\{ \int_y^a f(x, y)\, dx \right\} dy.$$

Proof. Consider the two functions of t, $0 \le t \le a$, defined by

$$\int_0^t \left\{ \int_0^x f(x, y)\, dy \right\} dx, \qquad \int_0^t \left\{ \int_y^t f(x, y)\, dx \right\} dy.$$

By the remarks following Theorem 7.1, the derivatives of these functions with respect to t are

$$\int_0^t f(t, y)\, dy, \qquad \int_0^t f(t, y)\, dy + \int_t^t f(x, t)\, dx.$$

Thus these functions differ by a constant. Since both are zero when $t = 0$, they are identical. □

Finally we need to discuss *polar coordinates*. If $(x, y) \in \mathbb{R}^2$ and $(x, y) \neq (0, 0)$, let

$$r = (x^2 + y^2)^{1/2}.$$

Then

$$(r^{-1}x)^2 + (r^{-1}y)^2 = 1,$$

so there is a unique θ, $0 \le \theta < 2\pi$, such that $\cos \theta = r^{-1}x$, $\sin \theta = r^{-1}y$. This means

$$x = r \cos \theta, \qquad y = r \sin \theta$$
$$r = (x^2 + y^2)^{1/2}, \qquad \theta = \tan^{-1}(y/x).$$

Thus any point p of the plane other than the origin is determined uniquely either by its *Cartesian coordinates* (x, y) or by its *polar coordinates* r, θ. A function defined on a subset of $\mathbb{R}^2$ can be expressed either as $f(x, y)$ or $g(r, \theta)$. These are related by

$$(7.4) \qquad f(x, y) = g((x^2 + y^2)^{1/2}, \tan^{-1}(y/x)),$$

$$(7.5) \qquad g(r, \theta) = f(r \cos \theta, r \sin \theta).$$

Theorem 7.4. *Suppose f is a continuous complex-valued function defined on the disc*

$$D_R = \{(x, y) \mid x^2 + y^2 < R\}.$$

Suppose g is related to f by (7.5). Then

$$\int_{-R}^{R} \left\{ \int_{-(R^2 - y^2)^{1/2}}^{(R^2 - y^2)^{1/2}} f(x, y)\, dx \right\} dy = \int_{0}^{R} \left\{ \int_{0}^{2\pi} g(r, \theta) r\, d\theta \right\} dr.$$

Proof. Look first at the quadrant $x \ge 0, y \ge 0$. For a fixed $y \ge 0$, if $x \ge 0$ then $x = (r^2 - y^2)^{1/2}$. Proposition 3.9 on coordinate changes gives

$$\int_{0}^{(R^2 - y^2)^{1/2}} f(x, y)\, dx = \int_{y}^{R} f((r^2 - y^2)^{1/2}, y)(r^2 - y^2)^{-1/2} r\, dr.$$

We integrate the integral on the right over $0 \le y \le R$, and use Theorem 7.3 to get

$$\int_{0}^{R} \left\{ \int_{0}^{r} f((r^2 - y^2)^{1/2}, y)(r^2 - y^2)^{-1/2}\, dy \right\} r\, dr.$$

Let $y = r \sin \theta$, $\theta \in [0, \tfrac{1}{2}\pi]$. Then $(r^2 - y^2)^{1/2} = r \cos \theta$. We may apply Proposition 3.9 to the preceding integral of $g \cdot r$ over $0 \le \theta \le \tfrac{1}{2}\pi$. Similar arguments apply to the other three quadrants. □

Exercises

1. Suppose $A \subset \mathbb{R}^2$ is open and suppose $g: A \to \mathbb{C}$ and $h: A \to \mathbb{C}$ are of class C^1. Show that a *necessary* condition for the existence of $f: A \to \mathbb{C}$ such that

$$(*) \qquad\qquad D_1 f = g, \qquad D_2 f = h$$

is that $D_2 g = D_1 h$.

2. In Exercise 1, suppose A is a disc $\{(x, y) \mid (x - x_0)^2 + (y - y_0)^2 < R^2\}$. Show that the condition $D_2 g = D_1 h$ is *sufficient*. (Hint: consider

$$f(x, y) = \int_{x_0}^{x} g(s, y) \, ds + \int_{y_0}^{y} h(x_0, t) \, dt.)$$

§8. Some infinitely differentiable functions

In §4 it was shown that any power series with a positive radius of convergence defines an infinitely differentiable function where it converges:

$$f(x) = \sum_{n=0}^{\infty} a_n(x - x_0)^n.$$

We know

$$a_n = (n!)^{-1} f^{(n)}(x_0).$$

In particular, if all derivatives of f are zero at x_0, then f is identically zero. There are infinitely differentiable functions which do not have this property.

Proposition 8.1. *There is an infinitely differentiable function $f: \mathbb{R} \to \mathbb{R}$ such that*

$$\begin{aligned} f(x) &= 0, & x \le 0, \\ f(x) &> 0, & x > 0. \end{aligned}$$

Proof. We define f by

$$\begin{aligned} f(x) &= 0, & x \le 0 \\ f(x) &= \exp(-1/x), & x > 0. \end{aligned}$$

Near any point $x \neq 0$, f is the composition of two infinitely differentiable functions. Repeated use of the chain rule shows that f is, therefore, infinitely differentiable except possibly at zero.

Let us show that f is continuous at 0. If $y > 0$, then

$$e^y = \sum_{n=0}^{\infty} (n!)^{-1} y^n > (m!)^{-1} y^m, \qquad m = 0, 1, \ldots.$$

Thus if $x > 0$,

$$0 < f(x) = \exp(-1/x) = \exp(1/x)^{-1} < m!(1/x)^{-m} = m! \, x^m,$$

$m = 0, 1, \ldots$. In particular, $f(x) \to 0$ as $x \to 0$.

It is easy to show by induction that for $x > 0$,

$$f^{(k)}(x) = p_k(x^{-1})f(x),$$

where $p_k(x)$ is a polynomial of degree $\leq k + 1$. Of course, this equation also holds for $x < 0$. Suppose we have shown that $f^{(k)}$ exists and is continuous at 0; then of course $f^{(k)}(0) = 0$. We have

$$(8.1) \qquad |[f^{(k)}(x) - f^{(k)}(0)]x^{-1}| = |f^{(k)}(x)x^{-1}|$$
$$= |x^{-1}p_k(x^{-1})f(x)|.$$

Since p_k is of order $\leq k + 1$ and

$$|f(x)| \leq (k + 3)! \, x^{k+3},$$

it follows that the right side of (8.1) converges to zero at $x \to 0$. Thus $f^{(k+1)}(0)$ exists and is zero. Similarly, $f^{(k+1)}(x) = p_{k+1}(x^{-1})f(x) \to 0$ as $x \to 0$, so $f^{(k+1)}$ is continuous. □

Note that all derivatives of the preceding function vanish at zero, but f is not identically zero. Therefore f does not have a convergent power series expansion around zero.

Corollary 8.2. *Suppose $a < b$. There is an infinitely differentiable function $g: \mathbb{R} \to \mathbb{R}$ such that*

$$g(x) = 0, \qquad x \notin (a, b),$$
$$g(x) > 0, \qquad x \in (a, b).$$

Proof. Let f be the function in Theorem 8.1 and let

$$g(x) = f(x - a)f(b - x).$$

This is positive where $x - a > 0$ and where $b - x > 0$, and is zero elsewhere. It is clearly of class C^∞. □

Corollary 8.3. *Suppose $a < b$. There is an infinitely differentiable function $h: \mathbb{R} \to \mathbb{R}$ such that*

$$h(x) = 0, \qquad x \leq a,$$
$$0 < h(x) < 1, \qquad a < x < b,$$
$$h(x) = 1, \qquad x \geq b.$$

Proof. Let g be the function in Corollary 8.2 and let

$$h(x) = c \int_{-\infty}^{x} g(t) \, dt,$$

where $c > 0$ is chosen so that $h(b) = 1$. Then $h' = cg \geq 0$, h is constant on $(-\infty, a]$ and on $[b, +\infty)$. □

Chapter 3

Periodic Functions and Periodic Distributions

§1. Continuous periodic functions

Suppose u is a complex-valued function defined on the real line $\mathbb{R}$. The function u is said to be *periodic* with *period* $a \neq 0$ if

$$u(x + a) = u(x)$$

for each $x \in \mathbb{R}$. If this is so, then also

$$u(x + 2a) = u((x + a) + a) = u(x + a) = u(x),$$
$$u(x - a) = u((x - a) + a) = u(x).$$

Thus u is also periodic with period $2a$ and with period $-a$. More generally, u is periodic with period na for each integer n. If u is periodic with period $a \neq 0$, then the function v,

$$v(x) = u(|a|x/2\pi)$$

is periodic with period 2π. It is convenient to choose a fixed period for our study of periodic functions, and the period 2π is particularly convenient. From now on the statement "u is periodic" will mean "u is periodic with period 2π." In this section we are concerned with continuous periodic functions. We denote the set of all continuous periodic functions from $\mathbb{R}$ to $\mathbb{C}$ by $\mathscr{C}$. This set includes, in particular, the functions

$$\sin nx, \qquad \cos nx, \qquad \exp(inx) = \cos nx + i \sin nx$$

for each integer n.

The set $\mathscr{C}$ can be considered a *vector space* in a natural way. We define the operations of addition and scalar multiplication by

$$(1) \qquad (u + v)(x) = u(x) + v(x), \qquad u, v \in \mathscr{C}, \qquad x \in \mathbb{R};$$

$$(2) \qquad (au)(x) = au(x), \qquad u \in \mathscr{C}, \qquad a \in \mathbb{C}, \qquad x \in \mathbb{R}.$$

it is easily checked that the functions $u + v$ and au are periodic. By Proposition 1.1 of Chapter 2, they are also continuous. Thus $u + v \in \mathscr{C}$, $au \in \mathscr{C}$. The axioms V1–V8 for a vector space are easily verified. We note also that there is a natural multiplication of elements of $\mathscr{C}$,

$$(uv)(x) = u(x)v(x), \qquad u, v \in \mathscr{C}, \qquad x \in \mathbb{R}.$$

The set $\mathscr{C}$ may also be considered as a *metric space*. Since the interval $[0, 2\pi]$ is a compact set in $\mathbb{R}$ and since $u \in \mathscr{C}$ is continuous,

$$\sup_{x \in [0, 2\pi]} |u(x)| < \infty.$$

69

We define the *norm* of $u \in \mathscr{C}$ to be the real number $|u|$ where

(3) $$|u| = \sup_{x \in \mathbb{R}} |u(x)| = \sup_{x \in [0, 2\pi]} |u(x)|.$$

The norm (3) has the following properties:

(4) $$|u| \geq 0, \quad \text{and} \quad |u| = 0 \quad \text{only if } u(x) = 0, \text{ all } x;$$

(5) $$|au| = |a| \, |u|, \qquad a \in \mathbb{C}, \qquad u \in \mathscr{C};$$

(6) $$|u + v| \leq |u| + |v|, \qquad u, v \in \mathscr{C}.$$

The properties (4) and (5) are easily checked. As for (6), suppose $x \in \mathbb{R}$. Then

$$|(u + v)(x)| = |u(x) + v(x)| \leq |u(x)| + |v(x)| \leq |u| + |v|.$$

Since this is true for every $x \in \mathbb{R}$, (6) is true.

To make $\mathscr{C}$ a metric space, we set

(7) $$d(u, v) = |u - v|.$$

Theorem 1.1. *The set $\mathscr{C}$ of continuous periodic functions is a vector space with the operations defined by* (1) *and* (2). *The set $\mathscr{C}$ is also a metric space with respect to the metric d defined by* (7), *and it is complete.*

Proof. As we noted above, checking that $\mathscr{C}$ satisfies the axioms for a vector space is straightforward. The axioms for a metric space are also easily checked, using (4), (5), and (6). For example,

$$d(u, w) = |u - w| = |(u - v) + (v - w)|$$
$$\leq |u - v| + |v - w| = d(u, v) + d(v, w).$$

Finally, suppose $(u_n)_{n=1}^{\infty}$ is a Cauchy sequence of functions in $\mathscr{C}$. By Theorem 4.1 of Chapter 2, there is a continuous function $u : \mathbb{R} \to \mathbb{C}$ such that $|u_n - u| \to 0$. Clearly u is periodic, so $u \in \mathscr{C}$ and $\mathscr{C}$ is complete. □

Sets which are simultaneously vector spaces and metric spaces of this sort are common enough and important enough to have been named and studied in the abstract. Suppose $\mathbf{X}$ is a real or complex vector space. A *norm* on $\mathbf{X}$ is a function assigning to each $\mathbf{u} \in \mathbf{X}$ a real number $|\mathbf{u}|$, such that

$$|\mathbf{u}| \geq 0, \quad \text{and} \quad |\mathbf{u}| = 0 \quad \text{implies} \quad \mathbf{u} = 0;$$
$$|a\mathbf{u}| = |a| \, |\mathbf{u}|, \qquad a \text{ scalar}, \mathbf{u} \in \mathbf{X};$$
$$|\mathbf{u} + \mathbf{v}| \leq |\mathbf{u}| + |\mathbf{v}|, \qquad \mathbf{u}, \mathbf{v} \in \mathbf{X}.$$

A *normed linear space* is a vector space $\mathbf{X}$ together with a norm $|\mathbf{u}|$. Associated to the norm is the metric

$$d(\mathbf{u}, \mathbf{v}) = |\mathbf{u} - \mathbf{v}|.$$

If the normed linear space is complete with respect to this metric, it is said to be a *Banach space*.

In this terminology, Theorem 1.1 has a very brief statement: $\mathscr{C}$ is a Banach space.

Suppose $\mathbf{X}$ is a complex normed linear space. A linear functional $F: \mathbf{X} \to \mathbf{C}$ is said to be *bounded* if there is a constant $c \geq 0$ such that

$$|F(\mathbf{u})| \leq c|\mathbf{u}|, \qquad \text{all } \mathbf{u} \in \mathbf{X}.$$

Proposition 1.2. *A linear functional F on a normed linear space X is continuous if and only if it is bounded.*

Proof. Suppose F is bounded. Then

$$|F(\mathbf{u}) - F(\mathbf{v})| = |F(\mathbf{u} - \mathbf{v})| \leq c|\mathbf{u} - \mathbf{v}|,$$

so $|F(\mathbf{u}) - F(\mathbf{v})| < \varepsilon$ if $|\mathbf{u} - \mathbf{v}| < c^{-1}\varepsilon$.

Conversely, suppose F is continuous. There is a $\delta > 0$ such that $|\mathbf{u}| = |\mathbf{u} - \mathbf{0}| \leq \delta$ implies

$$|F(\mathbf{u})| = |F(\mathbf{u}) - F(\mathbf{0})| \leq 1.$$

For any $\mathbf{u} \neq \mathbf{0}$, $\mathbf{u} \in X$, the vector $\mathbf{v} = \delta|\mathbf{u}|^{-1}\mathbf{u}$ has norm δ. Therefore

$$|F(\mathbf{u})| = |F(\delta^{-1}|\mathbf{u}|\mathbf{v})| = \delta^{-1}|\mathbf{u}| \, |F(\mathbf{v})| \leq \delta^{-1}|\mathbf{u}|,$$

and F is bounded. $\square$

It is important both in theory and practice to determine all the continuous linear functionals on a given space of functions. The reason is that many problems, in theory and in practice, can be interpreted as problems about existence or uniqueness of linear functionals satisfying given conditions. The examples below show that it is not obvious that there is any way to give a unified description of all the continuous linear functionals on $\mathscr{C}$. In fact one *can* give such a description (in terms of Riemann-Stieltjes integrals, or integrals with respect to a bounded Borel measure), but we shall not do this here. Instead we introduce a second useful space of periodic functions and determine the continuous linear functionals on this second space.

Exercises

1. Suppose $(a_n)_{n=-\infty}^{\infty}$ is a (two sided) sequence of complex numbers such that

$$\sum_{n=-\infty}^{\infty} |a_n| < \infty;$$

here we take the infinite sum to be

$$\sum_{n=-\infty}^{\infty} |a_n| = \sum_{n=0}^{\infty} |a_n| + \sum_{n=1}^{\infty} |a_{-n}|.$$

Show that the function u defined by

$$u(x) = \sum_{n=-\infty}^{\infty} a_n \exp{(inx)}$$

is continuous and periodic

2. Suppose $u: \mathbb{R} \to \mathbb{C}$ is a continuous function and suppose there is a constant M such that $u(x) = 0$ if $|x| \geq M$. Show that for any x the series

$$v(x) = \sum_{n=-\infty}^{\infty} u(x + 2n\pi)$$

converges. Show that the function v is in $\mathscr{C}$.

3. If $u \in \mathscr{C}$, define the real number $|u|'$ by

$$|u|' = (2\pi)^{-1} \int_0^{2\pi} |u(x)| \, dx.$$

Show that $|u|'$ is a norm on $\mathscr{C}$ and that $|u|' \leq |u|$.

4. Suppose d' is the metric associated with the norm $|u|'$ in Exercise 3. Show that $\mathscr{C}$ is *not* complete with respect to this metric. (Hint: take a sequence of functions $(u_n)_{n=1}^{\infty}$ of functions in $\mathscr{C}$ such that

$$0 \leq u_n(x) \leq 1, \qquad x \in \mathbb{R}, \, n = 1, 2, \ldots,$$
$$u_n(x) = 0, \qquad x \in [0, \pi/2 - 1/n] \cup [3\pi/2 + 1/n, 2\pi],$$
$$u_n(x) = 1, \qquad x \in [\pi/2, 3\pi/2].$$

Then $|u_n - u_m|' \to 0$ as $n, m \to \infty$. If $u \in \mathscr{C}$, there is an open interval $(\pi/2 - \delta, \pi/2 + \delta)$ on which either $|u(x)| > \frac{1}{3}$ or $|u(x) - 1| > \frac{1}{3}$. Show that $|u_n - u|' > \delta/6\pi$ for large values of n.)

5. Which of the following are bounded linear functionals on $\mathscr{C}$, with respect to the norm $|u|$?

(a) $F(u) = u(\pi/2)$,
(b) $F(u) = \int_0^{2\pi} \sin nx \, u(x) \, dx$,
(c) $F(u) = \int_0^{2\pi} (u(x))^2 \, dx$,
(d) $F(u) = 17u(0) + \int_\pi^{2\pi} u(x) \, dx$.
(e) $F(u) = -3|u(0)|$.

6. Suppose $\mathbf{X}$ is a normed linear space. Let $\mathbf{X}'$ be the set of all bounded linear functionals on $\mathbf{X}$. Then $\mathbf{X}'$ is a vector space. For $F \in \mathbf{X}'$, let

$$|F| = \sup \{|F(\mathbf{u})| \mid \mathbf{u} \in \mathbf{X}, |\mathbf{u}| \leq 1\}.$$

Show that $|F|$ is a norm on $\mathbf{X}'$. Show that for any $\mathbf{u} \in \mathbf{X}$ and $F \in \mathbf{X}'$,

$$|F(\mathbf{u})| \leq |F| \, |\mathbf{u}|.$$

Show that $\mathbf{X}'$ is a Banach space with respect to this norm.

§2. Smooth periodic functions

Suppose $u: \mathbb{R} \to \mathbb{C}$ is a continuous periodic function, and suppose that the derivative

$$Du(x) = u'(x)$$

exists for each $x \in \mathbb{R}$. Then Du is also periodic:

$$Du(x + 2\pi) = \lim_{h \to 0} h^{-1}[u(x + 2\pi + h) - u(x + 2\pi)]$$

$$= \lim_{h \to 0} h^{-1}[u(x + h) - u(x)] = Du(x).$$

In particular, if u is infinitely differentiable and periodic, then each derivative $Du, D^2u, \ldots, D^ku, \ldots$ is in $\mathscr{C}$.

We shall denote by $\mathscr{P}$ the subset of $\mathscr{C}$ which consists of all functions $u \in \mathscr{C}$ which are *smooth*, i.e., infinitely differentiable. Such a function will be called a *smooth periodic function*. If u is in $\mathscr{P}$, then the derivatives $Du, D^2u, \ldots$ are also in $\mathscr{P}$.

The set $\mathscr{P}$ is a subspace of $\mathscr{C}$ in the sense of vector spaces, so it is itself a vector space. The function $|\sin x|$ is in $\mathscr{C}$ but not in $\mathscr{P}$, so $\mathscr{P} \neq \mathscr{C}$. We could consider $\mathscr{P}$ as a metric space with respect to the metric on $\mathscr{C}$ given in the previous section, but we shall see later that $\mathscr{P}$ is not complete with respect to that metric. To be able to consider $\mathscr{P}$ as a complete space we shall introduce a new notion of convergence for functions in $\mathscr{P}$.

A sequence of functions $(u_n)_{n=1}^{\infty} \subset \mathscr{P}$ is said to *converge to $u \in \mathscr{P}$ in the sense of $\mathscr{P}$* if for each $k = 0, 1, 2, \ldots$,

$$|D^ku_n - D^ku| \to 0 \quad \text{as} \quad n \to \infty.$$

(Here $D^0u = u$.) We denote this by

$$u_n \to u \quad (\mathscr{P}).$$

Thus $(u_n)_{n=1}^{\infty}$ converges to u in the sense of $\mathscr{P}$ if and only if each derivative of u_n converges uniformly to the corresponding derivative of u as $n \to \infty$.

A sequence of functions $(u_n)_{n=1}^{\infty}$ is said to be *a Cauchy sequence in the sense of $\mathscr{P}$* if for each $k = 0, 1, \ldots, (D^ku_n)_{n=1}^{\infty}$ is a Cauchy sequence in $\mathscr{C}$. Thus

$$|D^ku_n - D^ku_m| \to 0 \quad \text{as} \quad n, m \to \infty$$

for each k.

When there is no danger of confusion we shall speak simply of "convergence" and of a "Cauchy sequence," without referring to the "the sense of $\mathscr{P}$." The statement of the following theorem is to be understood in this way.

Theorem 2.1. *The set $\mathscr{P}$ of all smooth periodic functions is a vector space. If $(u_n)_{n=1}^{\infty} \subset \mathscr{P}$ is a Cauchy sequence, then it converges to a function $u \in \mathscr{P}$.*

Proof. As noted above, $\mathscr{P}$ is a subspace of the vector space $\mathscr{C}$: if $u, v \in \mathscr{P}$ then $u + v \in \mathscr{P}$, $au \in \mathscr{P}$. Thus $\mathscr{P}$ is a vector space.

Suppose $(u_n)_{n=1}^{\infty}$ is a Cauchy sequence. For each k the sequence of derivatives $(D^ku_n)_{n=1}^{\infty}$ is a Cauchy sequence in $\mathscr{C}$. Therefore it converges uniformly to a function $v_k \in \mathscr{C}$. For $k = 0, 1, 2, \ldots$,

$$D^ku_n(x) = D^ku_n(0) + \int_0^x D^{k+1}u_n(t)\, dt.$$

By Theorem 4.2 of Chapter 2,

$$v_k(x) = \lim_{n \to \infty} D^k u_n(x) = \lim_{n \to \infty} D^k u_n(0) + \lim_{n \to \infty} \int_0^x D^{k+1} v_n(t)\, dt$$

$$= v_k(0) + \int_0^x v_{k+1}(t)\, dt.$$

Therefore $Dv_k = v_{k+1}$, all k. This means that if $u = v_0$, then $v_k = D^k u$ and $|D^k u_n - D^k u| \to 0$ as $n \to \infty$. Thus $u_n \to u$ (in the sense of $\mathscr{P}$). □

The remainder of this section is not necessary for the subsequent development. We show that there is no way of choosing a *norm* on $\mathscr{P}$ so that convergence as defined above is equivalent to convergence in the sense of the metric associated with the norm. However, there is a way of choosing a *metric* on $\mathscr{P}$ (not associated with a norm) such that convergence in the sense of $\mathscr{P}$ is equivalent to convergence in the sense of the metric. Finally, we introduce the abstract concept which is related to $\mathscr{P}$ in the way that the concept of "Banach space" is related to $\mathscr{C}$.

Suppose there were a norm $|u|'$ on $\mathscr{P}$ such that a sequence $(u_n)_{n=1}^\infty \subset \mathscr{P}$ converges in the sense of $\mathscr{P}$ to $u \in \mathscr{P}$ if and only if

$$|u_n - u|' \to 0.$$

Then there would be a constant M and an integer N such that

(1) $$|u|' \le M(|u| + |Du| + \cdots + |D^N u|), \qquad \text{all } u \in \mathscr{P}.$$

In fact, suppose (1) is false for every M, N. Then for each integer n there would be a $u_n \in \mathscr{P}$ such that

$$|u_n|' > n(|u_n| + |Du_n| + \cdots + |D^n u_n|).$$

Let

$$v_n = (|u_n|')^{-1} u_n.$$

Then

$$|D^k v_n| = (|u_n|')^{-1} |D^k u_n| < n^{-1} \qquad \text{if } n \ge k,$$

so $v_n \to 0$ in the sense of $\mathscr{P}$. But $|v_n|' = 1$, all n. This shows that the norm $|u|'$ must satisfy (1) for some M, N. Now let

$$w_n(x) = n^{-N-1} \sin nx, \qquad n = 1, 2, \ldots.$$

Then $w_n \in \mathscr{P}$ and

$$|D^k w_n| = n^{k-N-1} \le n^{-1}, \qquad k \le N.$$

Thus by (1),

$$|w_n|' \to 0,$$

But $|D^{N+1} w_n| = 1$, all n, so $(w_n)_{n=1}^\infty$ does not converge to 0 in the sense of $\mathscr{P}$. This contradicts our assumption about the norm $|u|'$.

Although we cannot choose a metric on $\mathscr{P}$, associated with a norm, which gives the right notion of convergence, we can choose a metric as follows. Let

$$d'(u, v) = \sum_{k=0}^{\infty} 2^{-k-1} |D^k u - D^k v| [1 + |D^k u - D^k v|]^{-1}.$$

The term of this sum indexed by k is non-negative and is smaller than 2^{-k-1}. Thus

$$d'(u, v) < 1, \qquad u, v \in \mathscr{P}.$$

It is clear that

$$d'(u, v) \geq 0, \qquad d'(u, v) = 0 \quad \text{implies} \quad u = v,$$
$$d'(u, v) = d'(v, u).$$

The triangle inequality is a little more difficult. Let

$$d(u, v) = |u - v|, \qquad d^*(u, v) = d(u, v)[1 + d(u, v)]^{-1}.$$

The reader may verify that

$$d^*(u, w) \leq d^*(u, v) + d^*(v, w).$$

Then

$$d'(u, w) = \sum_{k=0}^{\infty} 2^{-k-1} d^*(D^k u, D^k w)$$

$$\leq \cdots \leq d'(u, v) + d'(v, w).$$

Theorem 2.2. *A sequence of functions* $(u_n)_{n=1}^{\infty} \subset \mathscr{P}$ *is a Cauchy sequence in the sense of* $\mathscr{P}$ *if and only if it is a Cauchy sequence in the sense of the metric* d'. *Thus* $(\mathscr{P}, d')$ *is a complete metric space.*

Proof. Suppose $(u_n)_{n=1}^{\infty}$ is a Cauchy sequence in the sense of $\mathscr{P}$. Suppose $\varepsilon > 0$ is given. Choose k so large that $2^{-k} < \varepsilon$. Choose N so large that if $m \geq N$ and $n \geq N$, then

$$|D^j u_n - D^j u_m| < \tfrac{1}{2}\varepsilon, \qquad j = 0, 1, \ldots, k.$$

Then if $m, n \geq N$,

$$d'(u_m, u_n) = \sum_{j=0}^{\infty} 2^{-j-1} d^*(D^j u_m, D^j u_n)$$

$$< \sum_{j=0}^{k} 2^{-j-1}(\tfrac{1}{2}\varepsilon) + \sum_{j=k+1}^{\infty} 2^{-j-1}$$

$$< \tfrac{1}{2}\varepsilon + 2^{-k-1} < \tfrac{1}{2}\varepsilon + \tfrac{1}{2}\varepsilon.$$

Conversely, suppose $(u_n)_{n=1}^{\infty}$ is a Cauchy sequence in the sense of the metric d'. Given an integer $k \geq 0$ and an $\varepsilon > 0$, choose N so large that if $m, n \geq N$ then

$$d'(u_m, u_n) < 2^{-k-2}\varepsilon.$$

For $m, n \geq N$,

$$
\begin{aligned}
|D^k u_n - D^k u_m| &= 2^{k+2} \cdot 2^{-k-1} \cdot 2^{-1} |D^k u_n - D^k u_m| \\
&< 2^{k+2} \cdot 2^{-k-1} \cdot d^*(D^k u_n, D^k u_m) \\
&< 2^{k+2} d'(u_n, u_m) < \varepsilon.
\end{aligned}
$$

Thus $(u_n)_{n=1}^\infty$ is a Cauchy sequence in the sense of $\mathscr{P}$.

The same argument shows that $d'(u_n, u) \to 0$ if and only if $u_n \to u$ $(\mathscr{P})$. Thus $(\mathscr{P}, d')$ is complete. □

There is an important generalization of the concept of a Banach space, which includes spaces like $\mathscr{P}$. Let $\mathbf{X}$ be a vector space over the real or complex numbers. A *seminorm* on $\mathbf{X}$ is a function $\mathbf{u} \to |\mathbf{u}|$ from X to $\mathbb{R}$ such that

$$
|\mathbf{u}| \geq 0, \qquad |a\mathbf{u}| = |a| \, |\mathbf{u}|, \qquad |\mathbf{u} + \mathbf{v}| \leq |\mathbf{u}| + |\mathbf{v}|.
$$

(Thus a seminorm is a norm if and only if $|\mathbf{u}| = 0$ implies $\mathbf{u} = 0$.) Suppose there is given a sequence of seminorms on X, $|\mathbf{u}|_1, |\mathbf{u}|_2, \ldots$, with the property that

$$
(2) \qquad\qquad |\mathbf{u}|_k = 0, \qquad \text{all } k \text{ implies } \mathbf{u} = 0.
$$

Then we may define a metric on $\mathbf{X}$ by

$$
d'(\mathbf{u}, \mathbf{v}) = \sum_{k=1}^\infty 2^{-k} |\mathbf{u} - \mathbf{v}|_k [1 + |\mathbf{u} - \mathbf{v}|_k]^{-1}.
$$

If X is complete with respect to the metric d', it is said to be a *Frechet space*. Note that

$$
d'(\mathbf{u}_n, \mathbf{v}) \to 0 \quad \text{as} \quad n \to \infty
$$

is equivalent to

$$
|\mathbf{u}_n - \mathbf{v}|_k \to 0 \quad \text{as} \quad n \to \infty, \qquad \text{for all } k.
$$

In particular, if we take $\mathbf{X} = \mathscr{P}$ and

$$
|u|_k = |D^{k-1} u|,
$$

then d' agrees with d' as defined above. Thus Theorems 3.1 and 3.2 say that $\mathscr{P}$ is a Frechet space.

Exercises

1. Which of the following are Cauchy sequences in the sense of $\mathscr{P}$?

$$
\begin{aligned}
u_n(x) &= n^{-3} \cos nx, \\
v_n(x) &= (n!)^{-1} \sin nx, \\
w_n(x) &= \sum_{m=1}^n (m!)^{-1} \sin mx.
\end{aligned}
$$

2. Suppose X is a complex vector space with a sequence of seminorms $|u|_1, |u|_2, \ldots,$ satisfying (2). Let d' be the associated metric. Show that a linear functional $F: X \to \mathbb{C}$ is continuous if and only if there are a constant M and an integer N such that

$$|F(u)| \le M(|u|_1 + |u|_2 + \cdots + |u|_N), \qquad \text{all } u \in X.$$

§3. Translation, convolution, and approximation

The aim of this section and the next is to show that the space $\mathscr{P}$ of smooth periodic functions is dense in the space $\mathscr{C}$ of continuous periodic functions; in other words, any continuous periodic function u is the uniform limit of a sequence $(u_n)_{n=1}^{\infty}$ of smooth periodic functions. Even more important than this theorem is the method of proof, because we develop a systematic procedure for approximating functions by smooth functions.

The idea behind this procedure is that an average of translates of a function u is smoother than u itself, while if the translated functions are translated only slightly, the resulting functions are close to u. To illustrate this the reader is invited to graph the following functions from $\mathbb{R}$ to $\mathbb{R}$:

$$u_1(x) = |x|,$$
$$u_2(x) = \tfrac{1}{2}|x - \varepsilon| + \tfrac{1}{2}|x + \varepsilon|,$$
$$u_3(x) = \tfrac{1}{3}|x - \varepsilon| + \tfrac{1}{3}|x| + \tfrac{1}{3}|x + \varepsilon|,$$

where $\varepsilon > 0$.

If $u \in \mathscr{C}$ and $t \in \mathbb{R}$, the translation of u by t is the function $T_t u$,

$$T_t u(x) = u(x - t), \qquad x \in \mathbb{R}.$$

Then $T_t u \in \mathscr{C}$. The graph of T_t is the graph of u shifted t units to the right (i.e., shifted $|t|$ units to the left, if $t < 0$). In these terms the functions above are

$$u_2 = \tfrac{1}{2}T_\varepsilon u_1 + \tfrac{1}{2}T_{-\varepsilon}u_1, \qquad u_3 = \tfrac{1}{3}T_\varepsilon u_1 + \tfrac{1}{3}T_0 u_1 + \tfrac{1}{3}T_{-\varepsilon}u_1.$$

More generally, one could consider weighted averages of the form

(1) $$w = a_0 T_{t_0} u + a_1 T_{t_1} u + \cdots + a_r T_{t_r} u,$$

where

$$a_k > 0, \qquad a_0 + a_1 + \cdots + a_r = 1,$$

and most of the t_k are near 0. If

$$0 \le t_0 < t_1 < \cdots < t_k = 2\pi$$

and we set

$$b(t_k) = a_k(t_k - t_{k-1})^{-1}$$

then (1) becomes

(2) $$w = \sum b(t_k)(T_{t_k}u)(t_k - t_{k-1}).$$

The natural continuous analog of (2) is the symbolic integral

$$v = \int_0^{2\pi} b(t) T_t u \, dt,$$

defining the function

(3) $$v(x) = \int_0^{2\pi} b(t) u(x - t) \, dt.$$

We wish to study integrals of the form (3). If $u, v \in \mathscr{C}$, the *convolution* of u and v is the function $u * v$ defined by

(4) $$(u * v)(x) = \frac{1}{2\pi} \int_0^{2\pi} u(x - y) v(y) \, dy.$$

It follows readily from (4) that

(5) $$|u * v| \leq |u| \frac{1}{2\pi} \int_0^{2\pi} |v(x)| \, dx \leq |u| \, |v|.$$

Proposition 3.1. *If $u, v \in \mathscr{C}$, then $u * v \in \mathscr{C}$. Moreover*

(6) $$u * v = v * u,$$

(7) $$(au) * v = a(u * v), \qquad a \in \mathbb{C},$$

(8) $$(u + v) * w = u * w + v * w, \qquad w \in \mathscr{C},$$

(9) $$(u * v) * w = u * (v * w),$$

(10) $$T_t(u * v) = (T_t u) * v = u * (T_t v).$$

Proof. We begin with part of (10).

(11) $$T_t(u * v)(x) = (u * v)(x - t) = \frac{1}{2\pi} \int u(x - t - y) v(y) \, dy$$

$$= \frac{1}{2\pi} \int T_t u(x - y) v(y) \, dy = (T_t u) * v.$$

Therefore,

(12) $$|T_t(u * v) - u * v| = |(T_t u - u) * v|$$
$$\leq |T_t u - u| \, |v|,$$

where we have assumed (7) and (8). Now u is uniformly continuous on $[0, 2\pi]$ and is periodic; it follows easily that u is uniformly continuous on $\mathbb{R}$. Therefore $|T_t u - u| \to 0$ as $t \to 0$. Then (12) implies continuity of $u * v$. Also,

$$T_{2\pi}(u * v) = (T_{2\pi} u) * v = u * v,$$

so $u * v$ is periodic.

The equality of (6) follows from a change of variables in (4): let $y' = x - y$, and use the periodicity of u and v. Equalities (7), (8), and (9) are easy computations. The last part of (10) follows from (11) and (6). □

Note that

$$t^{-1}[u(t + x) - u(x)] = t^{-1}[T_{-t}u - u](x).$$

Lemma 3.2. *If $u \in \mathscr{P}$, then*

$$|t^{-1}(T_{-t}u - u) - Du| \to 0 \quad as \quad t \to 0.$$

Proof. By the Mean Value Theorem,

$$t^{-1}[T_{-t}u - u](x) - Du(x) = Du(y) - Du(x)$$

where $y = y(t, x)$ lies between x and $x + t$. Since Du is uniformly continuous, $t^{-1}(T_{-t}u - u)$ converges uniformly to Du as $t \to 0$. $\square$

Corollary 3.3. *If $u \in \mathscr{P}$, then*

$$t^{-1}(T_{-t}u - u) \to Du \ (\mathscr{P})$$

as $t \to 0$.

Proof. It is easy to see that

$$D^k(T_{-t}u) = T_{-t}(D^k u).$$

Then

$$D^k[t^{-1}(T_{-t}u - u)] = t^{-1}(T_{-t}D^k u - D^k u),$$

which converges uniformly to $D(D^k u) = D^k(Du)$. $\square$

Proposition 3.4. *If $u \in \mathscr{P}$ and $v \in \mathscr{C}$ then $u * v \in \mathscr{P}$ and*

(13) $$D^k(u * v) = (D^k u) * v, \qquad all \ k.$$

Proof. By Proposition 3.1,

(14) $$t^{-1}[T_{-t}(u * v) - (u * v)] = [t^{-1}(T_{-t}u - u)] * v.$$

By Lemma 3.2 and (5), the expression on the right in (14) converges uniformly to $(Du) * v$ as $t \to 0$. Thus

$$D(u * v) = (Du) * v,$$

and $u * v$ has a continuous derivative. But, $Du \in \mathscr{P}$, so we also have

$$D^2(u * v) = D((Du) * v) = (D^2 u) * v.$$

By induction, (13) holds for all k. $\square$

Corollary 3.5. *Suppose $(v_n)_1^\infty \subset \mathscr{C}$, $v \in \mathscr{C}$, and $|v_n - v| \to 0$. Then for each $u \in \mathscr{P}$,*

$$u * v_n \to u * v \ (\mathscr{P}).$$

Proof. For each k,

$$D^k(u * v_n - u * v) = D^k u * (v_n - v).$$

It follows from this and (5) that

$$|D^k(u * v_n - u * v)| \to 0,$$

all k. □

Having established the general properties of convolution, let us return to the question of approximation. Suppose $u \in \mathscr{C}$. If $(\varphi_n)_{n=1}^{\infty}$ is a sequence of functions in $\mathscr{P}$, then each function

$$u_n = \varphi_n * u$$

is smooth. Thinking of u_n as a weighted average of translates of u, we can expect u_n to be close to u if φ_n has average value 1 and is concentrated near 0 and 2π (as a function on $[0, 2\pi]$).

A sequence $(\varphi_n)_{n=1}^{\infty} \subset \mathscr{C}$ is said to be an *approximate identity* if

(i) $\varphi_n(x) \geq 0$, all n, x;
(ii) $1/2\pi \int_0^{2\pi} \varphi_n(x)\, dx = 1$, all n;
(iii) for each $0 < \delta < \pi$,

$$\int_{\delta}^{2\pi - \delta} \varphi_n(x)\, dx \to 0 \quad \text{as} \quad n \to \infty.$$

Theorem 3.6. *Suppose $(\varphi_n)_1^{\infty} \subset \mathscr{C}$ is an approximate identity. Then for each $u \in \mathscr{C}$,*

$$|\varphi_n * u - u| \to 0 \quad as \quad n \to \infty.$$

Moreover, if $u \in \mathscr{P}$ then

$$\varphi_n * u \to u \ (\mathscr{P}) \quad as \quad n \to \infty.$$

Proof. Since $(2\pi)^{-1} \int_0^{2\pi} \varphi_n(y)\, dy = 1$ and $\varphi_n * u = u * \varphi_n$, we have

$$2\pi|(\varphi_n * u)(x) - u(x)| = \left| \int_0^{2\pi} u(x - y)\varphi_n(y)\, dy - u(x) \int_0^{2\pi} \varphi_n(y)\, dy \right|$$

$$= \left| \int_0^{2\pi} [u(x - y) - u(x)]\varphi_n(y)\, dy \right|$$

$$\leq \left| \int_0^{\delta} \right| + \left| \int_{\delta}^{2\pi - \delta} \right| + \left| \int_{2\pi - \delta}^{2\pi} \right|$$

$$\leq \sup_{|s| \leq \delta} |T_s u - u| \left(\int_0^{\delta} \varphi_n + \int_{2\pi - \delta}^{2\pi} \varphi_n \right) + 2|u| \int_{\delta}^{2\pi - \delta} \varphi_n$$

$$\leq \sup_{|s| \leq \delta} |T_s u - u| + 2|u| \int_{\delta}^{2\pi - \delta} \varphi_n.$$

Given $\varepsilon > 0$, we first choose $\delta > 0$ so small that $\delta < \pi$ and

$$|T_s u - u| \leq \pi\varepsilon \quad \text{if } |s| \leq \delta.$$

Then choose N so large that

$$2|u| \int_{\delta}^{2\pi - \delta} \varphi_n < \pi\varepsilon, \qquad n \geq N.$$

If $n \geq N$, then

$$|\varphi_n * u - u| < \varepsilon.$$

This proves the first assertion. Now suppose $u \in \mathscr{P}$. For each k,

$$D^k(\varphi_n * u) = D^k(u * \varphi_n) = (D^k u) * \varphi_n = \varphi_n * (D^k u),$$

which converges uniformly to $D^k u$. Thus

$$\varphi_n * u \to u \quad (\mathscr{P}). \qquad \qquad \Box$$

In the next section we shall construct a sequence in $\mathscr{P}$ which is an approximate identity. It will follow, using Proposition 3.4 and Theorem 3.6, that $\mathscr{P}$ is dense in $\mathscr{C}$.

Exercises

1. Let $e_k(x) = \exp(ikx)$, $k = 0, \pm 1, \pm 2, \ldots$. These functions are in $\mathscr{P}$ (see §6 of Chapter 2). Suppose $u \in \mathscr{C}$. Show that

$$e_k * u = a_k e_k,$$

where

$$a_k = \frac{1}{2\pi} \int_0^{2\pi} e^{-iky} u(y) \, dy.$$

2. Show that $e_k * e_k = 1$, and $e_j * e_k = 0$ if $j \neq k$.

§4. The Weierstrass approximation theorems

A *trigonometric polynomial* is a function of the form

(1) $$\varphi(x) = \sum_{k=-n}^{n} a_k \exp(ikx),$$

where the coefficients a_k are in $\mathbb{C}$. The reason for the terminology is that for $k > 0$,

$$\exp(\pm ikx) = [\exp(\pm ix)]^k = (\cos x \pm i \sin x)^k.$$

Therefore any function of the form (1) can be written as a polynomial in the trigonometric functions $\cos x$ and $\sin x$. Conversely, recall that

$$\cos x = \tfrac{1}{2}[\exp(ix) + \exp(-ix)],$$

$$\sin x = \frac{1}{2i}[\exp(ix) - \exp(-ix)].$$

Therefore any polynomial in $\cos x$ and $\sin x$ can be written in the form (1).

Lemma 4.1. *There is a sequence* $(\varphi_n)_1^\infty$ *of trigonometric polynomials which is an approximate identity.*

Proof. We want to choose a non-negative trigonometric polynomial φ such that

$$\varphi(0) = \varphi(2\pi) = 1,$$
$$\varphi(x) < 1 \quad \text{for} \quad 0 < x < 2\pi.$$

Then successive powers of φ will take values at points near 0 and 2π which are relatively much greater than those taken at points between 0 and 2π. We may take

$$\varphi(x) = \tfrac{1}{2}(1 + \cos x)$$

and set

$$\varphi_n(x) = c_n(1 + \cos x)^n$$

where c_n is chosen so that

$$\int_0^{2\pi} \varphi_n(x)\, dx = 2\pi.$$

We need to show that for each $0 < \delta < \pi$,

(2) $$\int_\delta^{2\pi - \delta} \varphi_n(x)\, dx \to 0 \quad \text{as} \quad n \to \infty$$

There is a number r, $0 < r < 1$, such that

(3) $$1 + \cos x < r(1 + \cos y)$$

if

(4) $$x \in [\delta,\, 2\pi - \delta], \qquad y \in [0,\, \tfrac{1}{2}\delta].$$

Then (3) and (4) imply

$$\varphi_n(x) = c_n(1 + \cos x)^n \le r^n \varphi_n(y),$$

so

$$\tfrac{1}{2}\delta \varphi_n(x) \le r^n \int_0^{\frac{1}{2}\delta} \varphi_n(y)\, dy \le 2\pi r^n,$$

or

$$\varphi_n(x) \le 4\pi \delta^{-1} r^n, \qquad x \in [\delta,\, 2\pi - \delta]$$

Thus $\varphi_n \to 0$ uniformly on $[\delta,\, 2\pi - \delta]$. ☐

Lemma 4.2. *If* φ *is a trigonometric polynomial and* $u \in \mathscr{C}$, *then* $\varphi * u$ *is a trigonometric polynomial.*

Proof. This follows from Exercise 1 of the preceding section. ☐

Theorem 4.3. *The trigonometric polynomials are dense in the space* $\mathscr{C}$ *of continuous periodic functions, and in the space* $\mathscr{P}$ *of smooth periodic functions.*

That is, if $u \in \mathscr{C}$ and $v \in \mathscr{P}$, there are sequences $(u_n)_1^\infty$ and $(v_n)_1^\infty$ of trigonometric polynomials such that

$$|u_n - u| \to 0$$

and

$$v_n \to v \quad (\mathscr{P}).$$

Proof. Let $(\varphi_n)_1^\infty$ be a sequence of trigonometric polynomials which is an approximate identity, as in Lemma 4.1. Let

$$u_n = \varphi_n * u, \qquad v_n = \varphi_n * v.$$

By Lemma 4.2, the functions u_n and v_n are trigonometric polynomials. By Theorem 3.6, $u_n \to u$ uniformly and $v_n \to v$ in the sense of $\mathscr{P}$. $\square$

Note that if u, v are real-valued, then so are the sequences $(u_n)_1^\infty$, $(v_n)_1^\infty$ constructed here.

Corollary 4.4. *$\mathscr{P}$ is dense in $\mathscr{C}$.*

Theorem 4.3 is due to Weierstrass. There is a better-known approximation theorem, also due to Weierstrass, which can be deduced from Theorem 4.3.

Theorem 4.5. (Weierstrass polynomial approximation theorem). *Let u be a complex-valued continuous function defined on a closed interval $[a, b] \subset \mathbb{R}$. Then there is a sequence $(p_n)_1^\infty$ of polynomials which converges uniformly to u on the interval $[a, b]$.*

Proof. Suppose first that $[a, b] = [0, \pi]$. We can extend u so that it is a function in $\mathscr{C}$; for example, let $u(-x) = u(x)$, $x \in [0, \pi]$ and take the unique periodic extension of this function. Then there is a sequence $(u_n)_1^\infty$ of trigonometric polynomials converging uniformly to u. Now the partial sums of the power series

$$\sum (m!)^{-1}(ikx)^m = \exp(ikx)$$

converge to $\exp(ikx)$ uniformly on $[0, 2\pi]$. Therefore for each n, we may replace the functions $\exp(ikx)$ in the expression of the form (1) for u_n by partial sums, so as to obtain a polynomial p_n with

$$|p_n(x) - u_n(x)| < n^{-1}, \qquad x \in [0, 2\pi].$$

Then $p_n \to u$ uniformly on $[0, 2\pi]$.

In the case of an arbitrary interval $[a, b]$, let

$$v(x) = u(a + (b - a)x/\pi), \qquad x \in [0, \pi].$$

Then v is continuous on $[0, \pi]$, so there is a sequence $(q_n)_1^\infty$ of polynomials with $q_n \to v$ uniformly on $[0, \pi]$. Let

$$p_n(y) = q_n(\pi(y - a)/(b - a)), \qquad y \in [a, b].$$

Then p_n is also a polynomial and $p_n \to u$ uniformly on $[a, b]$. ☐

Exercises

1. Suppose $u \in \mathscr{C}$ and suppose that for each integer k,

$$\int_0^{2\pi} u(x) \exp (ikx) \, dx = 0.$$

Show that $u = 0$.

2. Suppose $u: [a, b] \to \mathbb{C}$ is continuous, and for each integer $n \geq 0$,

$$\int_a^b u(x)x^n \, dx = 0.$$

Show that $u = 0$.

§5. Periodic distributions

In general, a "distribution" is a continuous linear functional on some space of functions. A *periodic distribution* is a continuous linear functional on the space $\mathscr{P}$. Thus a periodic distribution is a mapping $F: \mathscr{P} \to \mathbb{C}$ such that

$$\begin{aligned} F(au) &= aF(u), & a \in \mathbb{C}, \quad u \in \mathscr{P}; \\ F(u + v) &= F(u) + F(v), & u, v \in \mathscr{P}; \\ F(u_n) &\to F(u) & \text{if } u_n \to u \ (\mathscr{P}). \end{aligned}$$

If v is a continuous periodic function defined on $[0, 2\pi]$, then we define a linear functional $F = F_v$ by

(1) $$F_v(u) = \frac{1}{2\pi} \int_0^{2\pi} v(x)u(x) \, dx, \qquad u \in \mathscr{C}.$$

Then $F_v: \mathscr{C} \to \mathbb{C}$ is linear, and

$$|F_v(u_n) - F_v(u)| \leq |v| \, |u_n - u|.$$

Therefore F_v is continuous on $\mathscr{C}$. Its restriction to the subspace $\mathscr{P}$ is a periodic distribution. We say that a periodic distribution F *is a function* if there is a $v \in \mathscr{C}$ such that $F = F_v$. If so, we may abuse notation and write $F = v$.

Note that different functions $v, w \in \mathscr{C}$ define different distributions. In fact, suppose $F_v = F_w$. Choose $(u_n)_1^\infty \subset \mathscr{P}$ such that $u_n \to w^* - v^*$ uniformly, where $w^*(x) = w(x)^*$, the complex conjugate. Then

$$0 = 2\pi(F_w(u_n) - F_v(u_n)) = \int_0^{2\pi} (w(x) - v(x))u_n(x) \, dx \to \int_0^{2\pi} |w(x) - v(x)|^2 \, dx,$$

so $w = v$.

Not every periodic distribution is a function. For example, let $\delta: \mathbb{C} \to \mathbb{C}$ be defined by

$$(2) \qquad\qquad \delta(u) = u(0), \qquad u \in \mathscr{C}.$$

Then the restriction of δ to $\mathscr{P}$ is a periodic distribution. It is called the δ-*distribution*, or *Dirac δ-distribution*. To see that it is not a function, let

$$u_n(x) = (\tfrac{1}{2} + \tfrac{1}{2} \cos x)^n.$$

Then $\delta(u_n) = 1$, all n. But $u_n(x) \to 0$ uniformly for $x \in [\varepsilon, 2\pi - \varepsilon]$, any $\varepsilon > 0$. Also $0 \le u_n(x) \le 1$, all x, n. It follows from this that for any $v \in \mathscr{C}$, $F_v(u_n) \to 0$. Thus $\delta \ne F_v$.

The set of all periodic distributions is denoted by $\mathscr{P}'$. We consider $\mathscr{P}'$ as a vector space in the usual way: if $F, G \in \mathscr{P}'$, $u \in \mathscr{P}$, $a \in \mathbb{C}$, then

$$(F + G)(u) = F(u) + G(u),$$
$$(aF)(u) = aF(u).$$

Note that if v, w are continuous periodic functions, then

$$F_v + F_w = F_{v+w}, \qquad F_{av} = aF_v.$$

A sequence $(F_n)_1^\infty \subseteq \mathscr{P}'$ is said to *converge* to $F \in \mathscr{P}'$ *in the sense of* $\mathscr{P}'$ if

$$F_n(u) \to F(u), \qquad \text{all } u \in \mathscr{P}.$$

We denote convergence in the sense of $\mathscr{P}'$ by

$$F_n \to F \;\; (\mathscr{P}'),$$

or simply by

$$F_n \to F$$

when it is understood in what sense convergence is understood.

We want to define operations of complex conjugation, reversal, translation, and differentiation for periodic distributions. For any such operation there is a standard procedure for extending the operation from functions to distributions. For example, if $v \in \mathscr{C}$, the complex conjugate function v^* is defined by

$$v^*(x) = v(x)^*.$$

Then

$$F_{v^*}(u) = \frac{1}{2\pi} \int_0^{2\pi} v(x)^* u(x)\, dx = \left(\frac{1}{2\pi} \int_0^{2\pi} v(x) u^*(x)\, dx \right)^* = (F_v(u^*))^*.$$

Then we *define F^** for an arbitrary $F \in \mathscr{P}'$ by

$$(3) \qquad\qquad F^*(u) = F(u^*)^*, \qquad u \in \mathscr{P}.$$

Similarly, if $v \in \mathscr{C}$ we define the reversed function $\tilde{v}$ by

$$\tilde{v}(x) = v(-x).$$

Then

$$F_{\tilde{v}}(u) = \frac{1}{2\pi} \int_0^{2\pi} v(-x)u(x)\, dx = \frac{1}{2\pi} \int_0^{2\pi} v(x)u(-x)\, dx = F_v(\tilde{u}).$$

We *define* $F^\sim$, $F \in \mathcal{P}'$, by

(4) $$F^\sim(u) = F(\tilde{u}), \qquad u \in \mathcal{P}.$$

If $v \in \mathscr{C}$ and $t \in \mathbb{R}$, recall that the translate $T_t v$ is defined by

$$T_t v(x) = v(x - t).$$

Then

$$\frac{1}{2\pi} \int_0^{2\pi} T_t v(x)u(x)\, dx = \frac{1}{2\pi} \int_0^{2\pi} v(x - t)u(x)\, dx$$

$$= \frac{1}{2\pi} \int_0^{2\pi} v(x)u(x + t)\, dx = F_v(T_{-t}u).$$

We *define* $T_t F$, $F \in \mathcal{P}'$, by

(5) $$T_t F(u) = F(T_{-t}u), \qquad u \in \mathcal{P}.$$

If $v \in \mathcal{P}$ and $u \in \mathcal{P}$, then integration by parts gives

$$\frac{1}{2\pi} \int_0^{2\pi} Dv(x)u(x)\, dx = -\frac{1}{2\pi} \int_0^{2\pi} v(x)Du(x)\, dx = -F_v(Du).$$

We *define* DF, $F \in \mathcal{P}'$, by

(6) $$(DF)(u) = -F(Du), \qquad u \in \mathcal{P}.$$

Then inductively,

(7) $$(D^k F)(u) = (-1)^k F(D^k u), \qquad u \in \mathcal{P}.$$

Each of the linear functionals so defined is a periodic distribution. For example, if $u_n \to u$ $(\mathcal{P})$ then $Du_n \to Du$ $(\mathcal{P})$. It follows that

$$(DF)(u_n) = -F(Du_n) \to -F(Du) = DF(u),$$

so the derivative DF is continuous. Similarly, F^*, $\tilde{F}$, $T_t F$, and $D^k F$ are in $\mathcal{P}'$.

In particular, let us take $F = \delta$. Then

(8) $$\delta = \delta^* = \tilde{\delta}$$

(9) $$T_t \delta(u) = u(t),$$

(10) $$D^k \delta(u) = (-1)^k D^k u(0), \qquad u \in \mathcal{P}.$$

Proposition 5.1. *The operations in $\mathcal{P}'$ defined by equations* (3), (4), (5), *and* (6) *are continuous, in the sense that if $F_n \to F$ $(\mathcal{P}')$ then*

$$F_n^* \to F^* \quad (\mathcal{P}'),$$
$$F_n^\sim \to F^\sim \quad (\mathcal{P}'),$$
$$T_t F_n \to T_t F \quad (\mathcal{P}'),$$
$$DF_n \to DF \quad (\mathcal{P}').$$

Proof. Each of these assertions follows trivially from the definitions. For example, if $u \in \mathscr{P}$ then

$$(DF_n)(u) = -F_n(Du) \to -F(Du) = DF(u).$$

Thus $DF_n \to DF$ $(\mathscr{P}')$, etc. ▯

Recall that if $u \in \mathscr{P}$ then Du is the limit of the "difference quotient" $t^{-1}(T_{-t}u - u)$.

Proposition 5.2. *If $F \in \mathscr{P}'$, then*

(11)
$$t^{-1}(T_{-t}F - F) \to DF \; (\mathscr{P}')$$

as $t \to 0$.

Proof. Suppose $u \in \mathscr{P}$. By definition,

(12) $t^{-1}(T_{-t}F - F)(u) = t^{-1}F(T_t u) - t^{-1}F(u) = -F(t^{-1}[u - T_t u]).$

Now

(13)
$$t^{-1}[u - T_t u](x) = t^{-1}[u(x) - u(x - t)].$$

An argument like that proving Lemma 3.2 and Corollary 3.3 shows that the expression in (13) converges to Du in the sense of $\mathscr{P}$ as $t \to 0$. From this fact and (12) we get (11). ▯

As an example,

$$t^{-1}(T_{-t}\delta - \delta)(u) = t^{-1}[u(-t) - u(0)] \to -Du(0) = (D\delta)(u).$$

The real and imaginary parts of a function $v \in \mathscr{C}$ can be defined by

$$\mathrm{Re}\, v = \tfrac{1}{2}(v + v^*),$$

$$\mathrm{Im}\, v = \frac{1}{2i}(v - v^*).$$

Similarly, we *define* the real and imaginary parts of a periodic distribution F by

$$\mathrm{Re}\, F = \tfrac{1}{2}(F + F^*),$$

$$\mathrm{Im}\, F = \frac{1}{2i}(F - F^*).$$

F is said to be *real* if $F = F^*$. A function $v \in \mathscr{C}$ is said to be *even* if $v(x) = v(-x)$, all x; it is said to be *odd* if $v(x) = -v(-x)$, all x. These conditions may be written

$$v = \tilde{v}, \qquad v = -\tilde{v}.$$

Similarly, we say a periodic distribution F is *even* if

$$F = F^{\sim};$$

we say F is *odd* if

$$F = -F^{\sim}.$$

Exercises

1. Which of the following define periodic distributions?

   (a) $F(u) = Du(1) - 3u(2\pi)$.
   (b) $F(u) = \int_0^{2\pi} (u(x))^2 \, dx$.
   (c) $F(u) = \int_0^{2\pi} u(x) \, dx$.
   (d) $F(u) = \int_0^\pi u(x)(1 + x)^7 \, dx$.
   (e) $F(u) = -\int_0^{2\pi} D^3 u(x)|\cos 2x| \, dx$.
   (f) $F(u) = \sum_{j=0}^n a_j D^j u(t_j)$.
   (g) $F(u) = \sum_{j=0}^\infty (j!)^{-1} D^j u(0)$.

2. Verify (8), (9), (10)

3. Express the distributions in parts (a) and (f) of Exercise 1 in terms of the δ-distribution and its translates and derivatives.

4. Compute DF when F is the distribution in part (c) or (d) of Exercise 1.

5. Show that Re F and Im F are real. Show that $F = \text{Re } F + i \text{ Im } F$.

6. Show that F real and u real, $u \in \mathscr{P}$, imply $F(u)$ is real.

7. Show that F even and u odd, $u \in \mathscr{P}$, imply $F(u) = 0$. Show that F odd and u even, $u \in \mathscr{P}$, imply $F(u) = 0$.

8. Show that any $F \in \mathscr{P}'$ can be written uniquely as $F = G + H$, where $G, H \in \mathscr{P}'$ and G is even, H is odd.

9. Suppose that $v \in \mathscr{C}$ is differentiable at each point of $\mathbb{R}$ and $Dv = w$ is in $\mathscr{C}$. Show that

$$D(F_v) = F_w;$$

in other words, if $F = v$, then $DF = Dv$.

10. Suppose v is a continuous complex-valued function defined on the interval $[0, 2\pi]$ and that $Dv = w$ is continuous on $(0, 2\pi)$ and bounded. Define $F_v \in \mathscr{P}'$ by

$$F_v(u) = \frac{1}{2\pi} \int_0^{2\pi} v(x)u(x) \, dx, \qquad u \in \mathscr{P}.$$

Show that

$$DF_v(u) = (v(0) - v(2\pi))u(0) + \frac{1}{2\pi} \int_0^{2\pi} w(x)u(x) \, dx.$$

In other words,

$$DF_v = F_w + [v(0) - v(2\pi)]\delta.$$

11. Let $v(x) = |\sin \tfrac{1}{2}x|$. Compute

$$(D^2 F_v)(u), \qquad u \in \mathscr{P}.$$

§6. Determining the periodic distributions

We know that any continuous periodic function v may be considered as a periodic distribution F_v. The derivatives $D^k F_v$ are also periodic distributions, though in general they are not (defined by) functions. It is natural to ask whether *all* periodic distributions are of the form $D^k F_v$, $v \in \mathscr{C}$. The answer is nearly yes.

Theorem 6.1. *Suppose F is a periodic distribution. Then there is an integer $k \geq 0$, a continuous periodic function v, and a constant function f, such that*

$$(1) \qquad\qquad F = D^k F_v + F_f.$$

The proof of this theorem will be given later in this section, after several other lemmas and theorems. First we need the notion of the *order* of a periodic distribution. A periodic distribution F is said to be *of order* k (k an integer ≥ 0) if there is a constant c such that

$$|F(u)| \leq c\{|u| + |Du| + \cdots + |D^k u|\}, \qquad \text{all } u \in \mathscr{P}.$$

For example, δ is of order 0. If $v \in \mathscr{C}$ then $D^k F_v$ is of order k. It is true, but not obvious, that any $F \in \mathscr{P}'$ is of order k for *some* integer $k \geq 0$.

Theorem 6.2. *If $F \in \mathscr{P}'$, then there is an integer $k \geq 0$ such that F is of order k.*

Proof. If F is not of order k, there is a function $u_k \in \mathscr{P}$ such that

$$|F(u_k)| \geq (k + 1)\{|u_k| + |Du_k| + \cdots + |D^k u_k|\}.$$

Let

$$v_k = (k + 1)^{-1}\{|u_k| + |Du_k| + \cdots + |D^k u_k|\}^{-1} u_k.$$

Then we have

$$(2) \qquad\qquad |F(v_k)| \geq 1,$$

while

$$(3) \qquad\qquad |v_k| + |Dv_k| + \cdots + |D^k v_k| \leq (k + 1)^{-1}.$$

Suppose now that F were not of order k for any $k \geq 0$. Then we could find a sequence $(v_k)_{k=1}^{\infty} \subset \mathscr{P}$ satisfying (2) and (3) for each k. But (3) implies

$$v_k \to 0 \ (\mathscr{P}).$$

Then (2) contradicts the continuity of F. Thus F must be of order k, some k. □

Lemma 6.3. *Suppose $F \in \mathscr{P}'$ is of order 0. Then there is a unique continuous linear functional $F_1 : \mathscr{C} \to \mathbb{C}$ such that*

$$F(u) = F_1(u), \qquad \text{all } u \in \mathscr{P}.$$

Proof. By assumption there is a constant c such that

$$|F(u)| \leq c|u|, \qquad u \in \mathscr{P}.$$

If $u \in \mathscr{C}$, there is a sequence $(u_n)_{n=1}^{\infty} \subset \mathscr{P}$ such that $u_n \to u$ uniformly. Then

$$|F(u_n) - F(u_m)| \le c|u_n - u_m| \to 0,$$

so $(F(u_n))_{n=1}^{\infty}$ is a Cauchy sequence. Let

(4) $F_1(u) = \lim F(u_n).$

We want to show that $F_1(u)$ is independent of the particular sequence used to approximate u. If $(v_n)_1^{\infty} \subset \mathscr{P}$ and $v_n \to u$ uniformly, then

$$|u_n - v_n| \to 0$$

so

$$|F(u_n) - F(v_n)| \le c|u_n - v_n| \to 0.$$

Thus

$$\lim F(u_n) = \lim F(v_n).$$

The functional $F_1 : \mathscr{C} \to \mathscr{C}$ defined by (4) is easily seen to be linear. It is continuous ($=$ bounded), because

$$|F(u)| = \lim |F(u_n)| \le c \lim |u_n| = c|u|.$$

Conversely, suppose $F_2 : \mathscr{C} \to \mathscr{C}$ is continuous and suppose $F_2(u) = F(u)$, all $u \in \mathscr{P}$. For any $u \in \mathscr{C}$, let $(u_n)_1^{\infty} \subset \mathscr{P}$ be such that $u_n \to u$ uniformly. Then

$$F_2(u) = \lim F_2(u_n) = \lim F(u_n) = F_1(u). \qquad \square$$

(The remainder of this section is not needed subsequently.)

Lemma 6.4. *Suppose $F \in \mathscr{P}'$ is of order 0, and suppose $F(w) = 0$ if w is a constant function. Then there is a function $v \in \mathscr{C}$ such that*

$$D^2 F_v = F.$$

Proof. Let us suppose first that $F = F_f$, where $f \in \mathscr{C}$. We shall try to find a *periodic* function v such that $D^2 v = f$. Then we must have

$$Dv(x) = Dv(0) + \int_0^x f(t)\, dt = a + \int_0^x f(t)\, dt,$$

where a is to be chosen so that v is periodic. We may require $v(0) = 0$. Then

$$v(x) = \int_0^x Dv(t)\, dt = \int_0^x \left[a + \int_0^t f(s)\, ds \right] dt$$

$$= ax + \int_0^x \int_0^t f(s)\, ds\, dt.$$

We use Theorem 7.3 of Chapter 2 to reverse the order of integration and get

$$v(x) = ax + \int_0^x f(s)(x - s)\, ds.$$

Let

$$(x - s)^+ = 0 \quad \text{if } x < s, \qquad (x - s)^+ = x - s \quad \text{if } x \geq s.$$

Then

(5) $$v(x) = ax + \int_0^{2\pi} f(s)(x - s)^+ \, ds.$$

By assumption on f,

$$\int_0^{2\pi} bf(s) \, ds = 0, \qquad b \in \mathbb{C}.$$

Now we want to choose a in (5) so that v is periodic. This will be true if $v(2\pi) = 0$, i.e.,

$$0 = 2\pi a + \int_0^{2\pi} (2\pi - s)f(s) \, ds = 2\pi a - \int_0^{2\pi} sf(s) \, ds.$$

Thus

$$a = \frac{1}{2\pi} \int_0^{2\pi} sf(s) \, ds$$

and

(6) $$v(x) = \frac{1}{2\pi} \int_0^{2\pi} f(s)[xs + 2\pi(x - s)^+] \, ds.$$

Now suppose only that $F \in \mathscr{P}'$ is of order 0 and that $F(w) = 0$ if w is constant. Let F_1 be the extension of F to a continuous linear functional on $\mathscr{C}$. Let

$$u_x(s) = xs + 2\pi(x - s)^+, \qquad 0 \leq s \leq 2\pi.$$

Then $u_x(0) = 2\pi x = u_x(2\pi)$. We can extend u_x so that it is a continuous periodic function of s. Then (6) suggests that we define a function v by

(7) $$v(x) = F_1(u_x), \qquad 0 \leq x \leq 2\pi.$$

We want to show that $v(0) = v(2\pi)$ and $D^2 F_v = F$. It is easy to check that

$$|u_x - u_y| \leq 2\pi|x - y|, \qquad u_{x+2\pi} = u_x.$$

Therefore

$$|v(x) - v(y)| \leq |F_1(u_x) - F_1(u_y)| \leq c|u_x - u_y| \leq c2\pi|x - y|,$$
$$v(2\pi) = F_1(u_{2\pi}) = F_1(4\pi^2) = 0 = v(0).$$

Thus v has an extension in $\mathscr{C}$. Let us compute DF_v. If $w \in \mathscr{P}$, then

$$2\pi t^{-1}[T_{-t}F_v - F_v](w) = t^{-1} \int_0^{2\pi} [v(x + t) - v(x)]w(x) \, dx.$$

Approximate the integral by Riemann sums. These give expressions of the form

(8) $\quad t^{-1} \sum [v(x_j + t) - v(x_j)]w(x_j)(x_j - x_{j-1})$

$$= F_1\left(\sum w(x_j)(x_j - x_{j-1})t^{-1} \cdot [u_{x_j+t} - u_{x_j}]\right),$$

since F_1 is linear. As partitions $(x_0, x_1, \ldots, x_n)$ of $(0, 2\pi)$ are taken with smaller mesh, the functions on which F_1 acts in (8) converge uniformly to the function g_t. Here

$$g_t(s) = \int_0^{2\pi} t^{-1}[u_{x+t}(s) - u_x(s)]w(x)\, dx.$$

Now $|t^{-1}(u_{x+t} - u_x)| \le 2\pi$. For fixed $s \in (0, 2\pi)$, and $0 < x < s$,

$$t^{-1}[u_{x+t}(s) - u_x(s)] \to s \quad \text{as} \quad t \to 0.$$

This convergence is uniform for x in any closed subinterval of $(0, s)$. Similarly,

$$t^{-1}[u_{x+t}(s) - u_x(s)] \to s + 2\pi \quad \text{as} \quad t \to 0,$$

uniformly for x in any closed subinterval of $(s, 2\pi)$. It follows that

(9) $\qquad 2\pi DF_v(w) = \lim_{t \to 0} t^{-1}[T_{-t}F_v - F_v](w) = F_1(g),$

where

$$g(s) = s \int_0^{2\pi} w(x)\, dx + 2\pi \int_s^{2\pi} w(x)\, dx.$$

Then

$$2\pi(D^2 F_v)(w) = -2\pi(DF_v)(Dw) = -F_1(h),$$

where

$$h(s) = s \int_0^{2\pi} Dw(x)\, dx + 2\pi \int_s^{2\pi} Dw(x)\, dx = 2\pi w(2\pi) - 2\pi w(s).$$

Since F_1 applied to a constant function gives zero, we have

$$2\pi(D^2 F_v)(w) = -F_1(h) = 2\pi F_1(w) = 2\pi F(w).$$

Thus $D^2 F_v = F$. $\quad\square$

Lemma 6.5. *Suppose $F \in \mathscr{P}'$ and suppose $F(w) = 0$ if w is a constant function. Then there is a unique $G \in \mathscr{P}'$ such that $DG = F$ and $G(w) = 0$ if w is a constant function. If F is of order $k \ge 1$, then G is of order $k - 1$.*

Proof. If $u \in \mathscr{P}$, it is not necessarily the derivative of a periodic function. We can get a periodic function by setting

$$Su(x) = \int_0^x u(t)\, dt - \frac{x}{2\pi} \int_0^{2\pi} u(t)\, dt = \int_0^x u(t)\, dt - xF_e(u),$$

where
$$e(x) = 1, \qquad \text{all } x,$$

Then
$$D(Su) = u - F_e(u)e.$$

It follows that if $DG = F$ and $G(e) = 0$, then

$$(10) \qquad G(u) = G(u - F_e(u)e) = G(D(Su)) = -DG(Su) = -F(Su).$$

Thus G is unique. To prove existence, we use (10) to *define* G. Since $S: \mathscr{P} \to \mathscr{P}$ is linear, G is linear. Also

$$|Su| \le 4\pi|u|,$$
$$|D(Su)| \le 2|u|,$$
$$|D^k(Su)| = |D^{k-1}u|, \qquad k \ge 2.$$

Then if $u_n \to u$ ($\mathscr{P}$) we have

$$G(u_n) = -F(Su_n) \to -F(Su) = G(u).$$

Thus $G \in \mathscr{P}'$. Also

$$DG(u) = -G(Du) = F(S(Du)) = F(u).$$

Finally, suppose F is of order $k \ge 1$. Then

$$|G(u)| = |F(Su)| \le c\{|Su| + |DSu| + \cdots + |D^k Su|\}$$
$$\le 5\pi c\{|u| + |Du| + \cdots + |D^{k-1}u|\},$$

and G is of order $k - 1$. $\square$

Corollary 6.6. *If $G \in \mathscr{P}'$ and $DG = 0$, then $G = F_f$, where f is constant.*

Proof. Again let $e(x) = 1$, all x. Let $f = G(e)e$, and

$$H = G - F_f.$$

Then $DH = 0$ and $H(e) = 0$. By Lemma 6.5 (uniqueness), $H = 0$. Thus $G = F_f$. $\square$

Finally, we can prove Theorem 6.1. Suppose $F \in \mathscr{P}'$. Take an integer k so large that F is of order $k - 2 \ge 0$. Again, let $e(x) = 1$, all x, $f = F(e)e$, and

$$F_0 = F - F_f.$$

Then F_0 is of order $k - 2$ and $F_0(e) = 0$. By repeated applications of Lemma 6.5 we can find $F_1, F_2, \ldots, F_{k-2} \in \mathscr{P}'$ so that

$$DF_j = F_{j-1}, \qquad F_j(e) = 0,$$

and F_j is of order $k - 2 - j$. Then F_{k-2} is of order 0. By Lemma 6.4, there is a $v \in \mathscr{C}$ such that $D^2 F_v = F_{k-2}$. Then

$$D^k F_v = D^{k-2} F_{k-2} = D^{k-1} F_{k-1} = \cdots = F_0$$
$$= F - F_f.$$

$\square$

Exercises

1. To what extent are the functions v and f in Theorem 6.1 uniquely determined?

2. Find $v \in \mathscr{C}$ such that $D^2F_v = F$, where

$$F = \delta - T_\pi\delta,$$

i.e.,

$$F(u) = u(0) - u(\pi).$$

3. Find $v \in \mathscr{C}$ and a constant function f such that

$$\delta = D^2F_v + F_f.$$

§7. Convolution of distributions

Suppose $v \in \mathscr{C}$ and $u \in \mathscr{P}$. The convolution $v * u$ can be written as

$$(v * u)(x) = (u * v)(x) = \frac{1}{2\pi} \int_0^{2\pi} u(x - y)v(y)\,dy$$

$$= \frac{1}{2\pi} \int_0^{2\pi} v(y)\tilde{u}(y - x)\,dy = F_v(T_x\tilde{u});$$

here again $\tilde{u}(x) = u(-x)$. Because of this it is natural to *define* the convolution of a periodic distribution F and a smooth periodic function u by the formula

(1) $(F * u)(x) = F(T_x\tilde{u}).$

Proposition 7.1.   *If F is a periodic distribution and u is a smooth periodic function, then the function $F * u$ defined by (1) is a smooth periodic function. Moreover,*

(2) $(aF) * u = a(F * u) = F * (au),$ $F \in \mathscr{P}', u \in \mathscr{P}, a \in \mathbb{C};$

(3) $(F + G) * u = F * u + G * u,$ $F \in \mathscr{P}', u \in \mathscr{P};$

(4) $F * (u + v) = F * u + F * v,$ $F \in \mathscr{P}', u, v \in \mathscr{P};$

(5) $T_t(F * u) = (T_tF) * u = F * (T_tu),$ $F \in \mathscr{P}', u \in \mathscr{P};$

(6) $D(F * u) = (DF) * u = F * (Du),$ $F \in \mathscr{P}', u \in \mathscr{P}.$

Proof.   The identities (2)–(5) follow from the definition (1) by elementary manipulations. For example,

$$T_t(F * u)(x) = (F * u)(x - t) = F(T_{x-t}\tilde{u})$$
$$= F(T_{-t}T_x\tilde{u}) = (T_tF)(T_x\tilde{u}) = ((T_tF) * u)(x).$$

Also,

$$F(T_{x-t}u) = F(T_x(T_tu)\tilde{\ }) = (F * (T_tu))(x).$$

This proves (5). It follows that

$$(F * u)(x + 2\pi) = (F * (T_{2\pi}u))(x) = (F * u)(x).$$

We know that

$$t^{-1}[T_{-t}F - F] \to DF \quad (\mathscr{P}') \quad \text{as} \quad t \to 0.$$

Therefore

$$t^{-1}[(F * u)(x + t) - (F * u)(x)] = t^{-1}[T_{-t}(F * u)(x) - (F * u)(x)]$$
$$= t^{-1}[T_{-t}F - F](T_x\tilde{u}) \to DF(T_x\tilde{u})$$
$$= ((DF) * u)(x).$$

This shows that $F * u$ is differentiable at each point $x \in \mathbb{R}$, with derivative $(DF) * u(x)$. By induction, $D^k(F * u) = (D^kF) * u$. Thus $F * u \in \mathscr{P}$. Finally, using (5) again,

$$t^{-1}[(F * u)(x + t) - (F * u)(x)] = F * [t^{-1}(T_{-t}u - u)](x)$$
$$= F(T_x[t^{-1}(T_{-t}u - u)]^\sim) \to F(T_x(Du)^\sim)$$
$$= (F * (Du))(x).$$

By induction, $D^k(F * u) = F * (D^ku)$, all k. We leave the proofs of (2), (3), (4) as an exercise. □

As an example:

$$(7) \qquad\qquad \delta * u = u, \qquad (D^k\delta) * u = D^ku.$$

In using (1) to define $F * u$, we departed from the procedure in §4, where operations on distributions were defined in terms of their actions on functions. Suppose $u \in \mathscr{C}$, $v \in \mathscr{P}$, $w \in \mathscr{P}$. Then

$$(8) \qquad F_{v*u}(w) = F_{u*v}(w) = (2\pi)^{-2} \int_0^{2\pi} \int_0^{2\pi} u(x - y)v(y)w(x) \, dy \, dx$$

$$= (2\pi)^{-2} \int_0^{2\pi} v(y) \left\{ \int_0^{2\pi} \tilde{u}(y - x)w(x) \, dx \right\} dy$$

$$= F_v(\tilde{u} * w).$$

This suggests that we could have defined $F * u$ as a *distribution* by letting it assign to $w \in \mathscr{P}$ the number $F(\tilde{u} * w)$. We shall see that this distribution corresponds to the function defined by (1).

Lemma 7.2. *If $u \in \mathscr{C}$ and $v \in \mathscr{C}$, then $w = u * v$ is the uniform limit of the functions w_n, where*

$$(9) \qquad\qquad w_n = \sum_{m=1}^n n^{-1}v(2\pi m/n)T_{2\pi m/n}u.$$

Proof. Let $x_{mn} = 2m\pi/n$. Then it is easy to see that

$$2\pi(w_n(x) - w(x)) = \sum_{m=1}^n \int_{x_{m-1,n}}^{x_{mn}} [v(x_{mn})u(x - x_{mn}) - v(y)u(x - y)] \, dy,$$

Now u, v are uniformly continuous and over the range of integration of the m-th summand,

$$|y - x_{mn}| < 2\pi/n.$$

Therefore $|w_n - w| \to 0$ as $n \to \infty$. $\quad \square$

Corollary 7.3. *If $u \in \mathscr{P}$ and $v \in \mathscr{C}$, then the functions w_n given by* (9) *converge to $w = u * v$ in the sense of $\mathscr{P}$ as $n \to \infty$.*

Proof. Since $D^k(T_x u) = T_x(D^k u)$, $D^k w_n$ is the corresponding sequence of functions for $(D^k u) * v = D^k(u * v)$. Therefore $D^k w_n \to D^k w$ uniformly as $n \to \infty$, for each k. $\quad \square$

Proposition 7.4. *If $F \in \mathscr{P}'$ and $u, v \in \mathscr{P}$, then*

$$F_f(v) = F(\tilde{u} * v),$$

where

$$f = F * u.$$

Proof. Let $w = \tilde{u} * v$ and let w_n be the corresponding function defined by (9), with $\tilde{u}$ replacing u. Then $w_n \to \tilde{u} * v$ $(\mathscr{P})$, so

$$F(\tilde{u} * v) = \lim F(w_n)$$

But

$$
\begin{aligned}
F(w_n) &= \sum_{m=1}^{n} n^{-1} v(2\pi m/n) F(T_{2\pi m/n} \tilde{u}) \\
&= \frac{1}{2\pi} \sum_{m=1}^{n} v(2\pi m/n) f(2\pi m/n) \frac{2\pi}{n} \\
&\to \frac{1}{2\pi} \int_0^{2\pi} v(x) f(x) \, dx = F_f(v). \qquad \square
\end{aligned}
$$

We shall now *define* the convolution of two periodic distributions F, G by

$$(10) \qquad\qquad (F * G)(u) = F(G^{\sim} * u), \qquad u \in \mathscr{P}.$$

If $G = F_v$, $f = F * v$, then Proposition 7.4 shows that $F_f = F * G$. In general, we must verify that (10) defines a periodic distribution. Clearly $F * G : \mathscr{P}' \to \mathbb{C}$ is linear. If $u_n \to u$ $(\mathscr{P})$ and G is of order k, then

$$
\begin{aligned}
|(G^{\sim} * u_n)(x) &- G^{\sim} * u(x)| \\
&= |(G^{\sim} * (u_n - u))(x)| \\
&= |G^{\sim}(T_x(u_n - u)^{\sim})| \\
&\le c\{|T_x(\tilde{u}_n - \tilde{u})| + |DT_x(\tilde{u}_n - \tilde{u})| + \cdots + |D^k T_x(\tilde{u}_n - \tilde{u})|\} \\
&= c\{|u_n - u| + |D(u_n - u)| + \cdots + |D^k(u_n - u)|\}.
\end{aligned}
$$

Thus $G^{\sim} * u_n \to G^{\sim} * u$ uniformly. Similarly, for each j, $D^j(G^{\sim} * u_n) = G^{\sim} * D_j u_n \to D^j(G^{\sim} * u)$ uniformly. Thus

$$G^{\sim} * u_n \to G^{\sim} * u \quad (\mathscr{P}),$$

so

$$(F * G)(u_n) \to (F * G)(u).$$

This shows that $F * G \in \mathscr{P}'$. As an example,

(11) $\delta * F = F * \delta = F.$

In the course of showing that $F * G$ is continuous, we have given an argument which proves the following.

Lemma 7.5. *If* $F \in \mathscr{P}'$, $(u_n)_1^\infty \subset \mathscr{P}$, *and* $u_n \to u$ $(\mathscr{P})$, *then* $F * u_n \to F * u$ $(\mathscr{P})$.

Corollary 7.6. *Suppose* $(f_n)_1^\infty \subset \mathscr{P}$, $(g_n)_1^\infty \subset \mathscr{P}$, *and set*

$$F_n = F_{f_n}, \qquad G_n = F_{g_n}.$$

Suppose

$$F_n \to F \ (\mathscr{P}') \quad and \quad G_n \to G \ (\mathscr{P}').$$

Then

$$F_n * G \to F * G \ (\mathscr{P}')$$

and

$$F * G_n \to F * G \ (\mathscr{P}').$$

Proof. Suppose $u \in \mathscr{P}$. Then

$$(F_n * G)(u) = F_n(G^\sim * u) \to F(G^\sim * u) = (F * G)(u)$$

Also, $G_n^\sim * u \to G^\sim * u$ $(\mathscr{P})$ so

$$(F * G_n)(u) = F(G_n^\sim * u) \to F(G^\sim * u) = (F * G)(u). \qquad \square$$

We can now prove approximation theorems for periodic distributions analogous to those for functions.

Theorem 7.7. *Suppose* $(\varphi_n)_1^\infty \subset \mathscr{P}$ *is an approximate identity, and suppose* $F \in \mathscr{P}'$. *Let* $F_n = F_{f_n}$, *where* $f_n = F * \varphi_n$. *Then* $F_n \to F$ $(\mathscr{P}')$.

In particular, there is a sequence $(f_n)_1^\infty$ *of trigonometric polynomials such that* $F_{f_n} \to F$ $(\mathscr{P}')$.

Proof. We have, by Proposition 7.4,

$$F_n(u) = F(\tilde{\varphi}_n * u), \qquad u \in \mathscr{P}.$$

But $(\tilde{\varphi}_n)_{n=1}^\infty$ is also an approximate identity, so

$$\tilde{\varphi}_n * u \to u \ (\mathscr{P}).$$

Therefore

$$F_n(u) \to F(u).$$

If $(\varphi_n)_{n=1}^\infty$ is an approximate identity consisting of trigonometric polynomials, then the functions $f_n = F * \varphi_n$ are also trigonometric polynomials. In fact, let

$$e_k(x) = \exp(ikx).$$

Then

$$(F * e_k)(x) = F(T_x \tilde{e}_k).$$

But

$$(T_x \tilde{e}_k)(y) = e_k(x - y) = e_k(x)\tilde{e}_k(y)$$

so

$$(F * e_k)(x) = F(\tilde{e}_k)e_k(x).$$

Thus

$$F * \left(\sum a_k e_k \right) = \sum a_k F(\tilde{e}_k) e_k$$

is a trigonometric polynomial. ☐

Finally, we prove the analog of Proposition 7.1 and Proposition 3.1.

Proposition 7.8. *Suppose* $F, G, H \in \mathscr{P}'$, $a \in \mathbb{C}$. *Then*

(12) $F * G = G * F,$

(13) $(aF) * G = a(F * G) = F * (aG),$

(14) $(F + G) * H = F * H + G * H,$

(15) $(F * G) * H = F * (G * H),$

(16) $T_t(F * G) = (T_t F) * G = F * (T_t G),$

(17) $D^k(F * G) = (D^k F) * G = F * (D^k G).$

Proof. All of these identities except (12) and (15) follow from the definitions by a sequence of elementary manipulations. As an example, we shall prove part of (16):

$$[T_t(F * G)](u) = F * G(T_{-t}u) = F(G^\sim * T_{-t}u)$$
$$= F((T_{-t}G^\sim) * u)$$
$$= F((T_t G)^\sim * u) = (F * T_t G)(u).$$

Here we used the identity

(18) $(T_t G)^\sim = T_{-t}(G^\sim).$

To prove (12) and (15) we use Theorem 7.7 and Corollary 7.6. First, suppose $G = F_g$, $g \in \mathscr{P}$. Take $(f_n)_1^\infty \subset \mathscr{P}$ such that

$$F_n = F_{f_n} \to F \quad (\mathscr{P}').$$

Let

$$h_n = f_n * g, \qquad H_n = F_{h_n}.$$

It follows from (8), (10), and Corollary 7.6 that

$$H_n = F_n * G \to F * G \quad (\mathscr{P}').$$

But

$$h_n = g * f_n,$$

so also

$$H_n = G * F_n \to G * F \quad (\mathscr{P}').$$

Thus (12) is true when $G = F_g$, $g \in \mathscr{P}$. In the general case, take $(g_n)_1^\infty \subset \mathscr{P}$ so that

$$G_n = F_{g_n} \to G.$$

Then, in the sense of $\mathscr{P}'$,

$$F * G = \lim F * G_n = \lim G_n * F = G * F.$$

The proof of (15) is similar. In the first place, (15) is true when

$$F = F_f, \qquad G = F_g, \qquad H = F_h,$$

since

$$F * (G * H) = F * F_{g*h} = F_{f*(g*h)} = F_{(f*g)*h} = (F * G) * H.$$

We then approximate an arbitrary F by F_{f_n} and get (15) when $G = F_g$, $H = F_h$. Then approximate G, H successively to get (15) for all F, G, $H \in \mathscr{P}'$. The rest of the proof is left as an exercise.

Exercises

1. Prove the identities (2), (3), (4).
2. Prove the identities (7), (11).
3. Prove the identities (13), (14), (16), (17), (18) directly from the definitions.
4. Prove the identities in Exercise 3 by approximating the distributions F, G, H by smooth periodic functions.

§8. Summary of operations on periodic distributions

In this section we simply collect for reference the definitions and results concerning $\mathscr{P}'$. The space $\mathscr{P}$ is the set of infinitely differentiable periodic functions $u: \mathbb{R} \to \mathbb{C}$. We say

$$u_n \to u \; (\mathscr{P}) \quad \text{if} \quad |D^k u_n - D^k u| \to 0, \qquad \text{all } k.$$

A periodic distribution is a mapping $F: \mathscr{P} \to \mathbb{C}$ with

$$F(u + v) = F(u) + F(v), \qquad F(au) = aF(u),$$
$$F(u_n) \to F(u) \quad \text{if} \quad u_n \to u \ (\mathscr{P}).$$

If $v \in \mathscr{C}$, the space of continuous periodic functions, then $F_v \in \mathscr{P}'$ is defined by

$$F_v(u) = \frac{1}{2\pi} \int_0^{2\pi} v(x)u(x) \, dx.$$

The δ-distribution is defined by

$$\delta(u) = u(0).$$

The sum, scalar multiple, complex conjugate, reversal, and translation of distributions are defined by

$$(F + G)(u) = F(u) + G(u),$$
$$(aF)(u) = aF(u),$$
$$F^*(u) = (F(u^*))^* \qquad (u^*(x) = u(x)^*),$$
$$F^\sim(u) = F(\tilde{u}) \qquad (\tilde{u}(x) = u(-x)),$$
$$(T_t F)(u) = F(T_{-t}u) \qquad (T_t u(x) = u(x - t)).$$

Derivatives are defined by

$$(D^k F)(u) = (-1)^k F(D^k u).$$

We say

$$F_n \to F \ (\mathscr{P}')$$

if

$$F_n(u) \to F(u), \qquad \text{all } u \in \mathscr{P}.$$

In particular,

$$t^{-1}(T_{-t}F - F) \to DF \quad \text{as} \quad t \to 0.$$

Then

$$\delta = \delta^* = \delta^\sim$$
$$(T_t\delta)(u) = u(t),$$
$$(D^k\delta)(u) = (-1)^k D^k u(0).$$

If $v \in \mathscr{C}$ then

$$(F_v)^* = F_{v^*},$$
$$(F_v)^\sim = F_{\tilde{v}},$$
$$T_t(F_v) = F_w, \quad \text{where} \quad w = T_t v.$$

If $v \in \mathscr{P}$,

$$D^k(F_v) = F_w \quad \text{where} \quad w = D^k v.$$

The convolution $F * u$ is the *function*

$$(F * u)(x) = F(T_x\tilde{u}), \qquad u \in \mathscr{P}.$$

Then $F * u \in \mathscr{P}$. If $v \in \mathscr{C}$, then

$$F_v * u = v * u.$$

If $(\varphi_n)_1^\infty \subset \mathscr{P}$ is an approximate identity, then

$$F * \varphi_n \to F \quad (\mathscr{P}').$$

More precisely,

$$F_{f_n} \to F \ (\mathscr{P}'), \quad \text{where} \quad f_n = F * \varphi_n.$$

In particular,

$$\delta * u = u, \qquad u \in \mathscr{P}.$$

The convolution $F * G$ is the *distribution*

$$(F * G)(u) = F(G^\sim * u), \qquad u \in \mathscr{P}.$$

In particular if $v \in \mathscr{P}$, then

$$F * F_v = F_f, \quad \text{where} \quad f = F * v.$$

Clearly

$$\delta * F = F * \delta = F.$$

If

$$F_n \to F \ (\mathscr{P}')$$

then

$$F_n * G \to F * G \ (\mathscr{P}').$$

The convolution of distributions satisfies

$$F * G = G * F,$$
$$(aF) * G = a(F * G) = F * (aG),$$
$$(F + G) * H = F * H + G * H,$$
$$(F * G) * H = F * (G * H),$$
$$T_t(F * G) = (T_tF) * G = F * (T_tG),$$
$$D^k(F * G) = (D^kF) * G = F * (D^kG).$$

A periodic distribution F is *real* if $F = F^*$. Any $F \in \mathscr{P}'$ can be written uniquely as

$$F = G + iH, \qquad G, H \text{ real}.$$

In fact

$$G = \operatorname{Re} F = \tfrac{1}{2}(F + F^*),$$

$$H = \operatorname{Im} F = \frac{1}{2i}(F - F^*).$$

A periodic distribution F is *even* if $F = F^\sim$ and *odd* if $F = -F^\sim$. Any $F \in \mathscr{P}'$ can be written uniquely as

$$F = G + H, \qquad G \text{ even, } H \text{ odd}.$$

In fact

$$G = \tfrac{1}{2}(F + F^{\sim}),$$
$$H = \tfrac{1}{2}(F - F^{\sim}).$$

The δ-distribution is real and even.

A periodic distribution is of *order k* if there is a constant c such that

$$|F(u)| \le c(|u| + |Du| + \cdots + |D^k u|), \qquad \text{all } u \in \mathscr{P}.$$

Any $F \in \mathscr{P}'$ is of order k for some k.

If $v \in \mathscr{C}$ and f is constant,

$$D^k F_v + F_f \in \mathscr{P}'.$$

Conversely, if $F \in \mathscr{P}'$ is of order $k - 2$, $k \ge 2$, then there are $v \in \mathscr{C}$ and constant function f such that

$$F = D^k F_v + F_f.$$

Chapter 4

Hilbert Spaces and Fourier Series

§1. An inner product in $\mathscr{C}$, and the space L^2

Suppose u and v are in $\mathscr{C}$, the space of continuous complex-valued periodic functions. The *inner product* of u and v is the number (u, v) defined by

$$(1) \qquad (u, v) = \frac{1}{2\pi} \int_0^{2\pi} u(x)v(x)^* \, dx.$$

It is easy to verify the following properties of the inner product:

$$(2) \qquad (au, v) = a(u, v) = (u, a^*v),$$

$$(3) \qquad (u_1 + u_2, v) = (u_1, v) + (u_2, v),$$

$$(4) \qquad (u, v_1 + v_2) = (u, v_1) + (u, v_2),$$

$$(5) \qquad (v, u) = (u, v)^*,$$

$$(6) \qquad (u, u) \geq 0, \qquad (u, u) = 0 \text{ only if } u = 0.$$

We define $\|u\|$ for $u \in \mathscr{C}$ by

$$(7) \qquad \|u\| = (u, u)^{1/2} = \left(\frac{1}{2\pi} \int_0^{2\pi} |u(x)|^2 \, dx \right)^{1/2}.$$

Lemma 1.1. *If $u, v \in \mathscr{C}$, then*

$$(8) \qquad |(u, v)| \leq \|u\| \, \|v\|.$$

Proof. If $v = 0$ then

$$(u, v) = (u, 0v) = 0(u, v) = 0,$$

and (8) is true. Suppose $v \neq 0$. Note that for any complex number a,

$$(9) \quad 0 \leq (u - av, u - av) = (u, u) - (av, u) - (u, av) + (av, av)$$
$$= \|u\|^2 - a(u, v)^* - a^*(u, v) + |a|^2 \|v\|^2.$$

Let

$$a = (u, v)\|v\|^{-2}.$$

Then (9) becomes

$$0 \leq \|u\|^2 - 2|(u, v)|^2\|v\|^{-2} + |(u, v)|^2\|v\|^{-2},$$

and this implies (8). $\square$

The inequality (8) is known as the *Schwarz inequality*. Note that only the properties (2)–(6) were used in the proof, and no other features of the inner product (1).

Corollary 1.2. *The function $u \to \|u\|$ is a norm on $\mathscr{C}$.*

Proof. Recall that this means that $\|u\|$ satisfies

(10) $\|u\| \geq 0,$ $\|u\| = 0$ only if $u = 0,$

(11) $\|au\| = |a|\,\|u\|,$ $a \in \mathbb{C},$

(12) $\|u + v\| \leq \|u\| + \|v\|.$

Property (10) follows from (6) and property (11) follows from (2). To prove (12), we take the square and use the Schwarz inequality:

$$\begin{aligned}
\|u + v\|^2 = (u + v, u + v) &= \|u\|^2 + (u, v) + (v, u) + \|v\|^2 \\
&\leq \|u\|^2 + 2\|u\|\,\|v\| + \|v\|^2 \\
&= (\|u\| + \|v\|)^2.
\end{aligned}$$

The new norm on $\mathscr{C}$ is dominated by the preceding norm:

(13) $\|u\| \leq |u| = \sup\,\{|u(x)|\}.$

It is important to note that $\mathscr{C}$ is *not* complete with respect to the metric associated with this new norm. For example, let $u_n : \mathbb{R} \to \mathbb{R}$ be the periodic function whose graph contains the line segments joining the pairs of points

$$(0, 0), \qquad \left(\tfrac{1}{2}\pi, 0\right);$$

$$\left(\tfrac{1}{2}\pi, 0\right), \qquad \left(\tfrac{1}{2}\pi + \tfrac{1}{n}, 1\right);$$

$$\left(\tfrac{1}{2}\pi + \tfrac{1}{n}, 1\right), \qquad \left(\tfrac{3}{2}\pi - \tfrac{1}{n}, 1\right);$$

$$\left(\tfrac{3}{2}\pi - \tfrac{1}{n}, 1\right), \qquad \left(\tfrac{3}{2}\pi, 0\right);$$

$$\left(\tfrac{3}{2}\pi, 0\right), \qquad (2\pi, 0).$$

Then for $m \geq n,$

$$\|u_n - u_m\|^2 = \frac{1}{2\pi} \int_0^{2\pi} |u_n - u_m|^2 \leq \frac{1}{2\pi} \cdot 2 \cdot \frac{1}{n} = \frac{1}{n\pi},$$

so $(u_n)_1^\infty$ is a Cauchy sequence in the new metric. However, there is no $u \in \mathscr{C}$ such that

$$\|u_n - u\| \to 0.$$

In order to get a complete space which contains $\mathscr{C}$ with this inner product, we turn to the space of periodic distributions. Suppose $(u_n)_1^\infty \subset \mathscr{C}$ is a Cauchy sequence with respect to the metric induced by the norm $\|u\|$, i.e.

$$\|u_n - u_m\| \to 0 \quad \text{as} \quad n, m \to \infty.$$

Let $(F_n)_1^\infty$ be the corresponding sequence of *distributions*:

$$F_n = F_{u_n}.$$

Thus if $v \in \mathscr{P}$,

$$F_n(v) = \frac{1}{2\pi} \int_0^{2\pi} u_n(x)v(x)\,dx = (v, u_n^*),$$

where again u_n^* denotes the complex conjugate function. By the Schwarz inequality,

$$\begin{aligned}|F_n(v) - F_m(v)| = |(v, u_n^* - u_m^*)| &\le \|v\|\,\|u_n - u_m\| \\ &\le |v|\,\|u_n - u_m\|.\end{aligned}$$

Therefore $(F_n(v))_1^\infty$ has a limit. We define

(14) $$F(v) = \lim F_n(v).$$

The functional $F: \mathscr{P} \to \mathbb{C}$ defined by (14) is clearly linear, since each F_n is linear. In fact, F is a periodic distribution. To see this, we take N so large that

$$\|u_n - u_m\| \le 1 \qquad \text{if } n, m \ge N.$$

Let

$$M = \max\{\|u_1\|, \|u_2\|, \ldots, \|u_N\|\} + 1.$$

Then for any $n \le N$,

$$\|u_n\| \le M,$$

while if $n > N$,

$$\begin{aligned}\|u_n\| = \|u_n - u_N + u_N\| &\le \|u_n - u_N\| + \|u_N\| \\ &< 1 + \|u_N\| \le M.\end{aligned}$$

Therefore

$$|F_n(v)| = |(v, u_n^*)| \le \|v\|\,\|u_n\| \le M|v|,$$

so

$$|F(v)| = \lim |F_n(v)| \le M|v|.$$

We have proved the following lemma.

Lemma 1.2. *If $(u_n)_1^\infty \subset \mathscr{C}$ is a Cauchy sequence with respect to the norm $\|u\|$, then the corresponding sequence of distributions*

$$F_n = F_{u_n}$$

converges in the sense of $\mathscr{P}'$ to a distribution F, which is of order 0.

It is important to know when two Cauchy sequences in $\mathscr{C}$ give rise to the same distribution.

Lemma 1.3. *Suppose $(u_n)_1^\infty \subset \mathscr{C}$ and $(v_n)_1^\infty \subset \mathscr{C}$ are Cauchy sequences with respect to the norm $\|u\|$. Let $F_n = F_{u_n}$ and $G_n = F_{v_n}$ be the corresponding distributions, and let F, G be the limits:*

$$F_n \to F \ (\mathscr{P}') \quad \text{and} \quad G_n \to G \ (\mathscr{P}').$$

Then $F = G$ if and only if

$$\|u_n - v_n\| \to 0.$$

Proof. Let $w_n = u_n - v_n$ and let $H_n = F_n - G_n = F_{w_n}$. We want to show

$$H_n \to 0 \; (\mathscr{P}') \qquad \text{if and only if } \|w_n\| \to 0.$$

Suppose

$$\|w_n\| \to 0.$$

Then for any $u \in \mathscr{P}$,

$$|H_n(u)| = |(u, w_n^*)| \leq \|u\| \, \|w_n\| \to 0.$$

Conversely, suppose

(15) $$H_n \to 0 \; (\mathscr{P}').$$

Given $\varepsilon > 0$, take N so large that $n, m \geq N$ implies

(16) $$\|w_n - w_m\| = \|(u_n - u_m) + (v_n - v_m)\| \leq \varepsilon.$$

Fix, $m \geq N$. Then if $n \geq N$ we use (16) to get

$$\begin{aligned}
\|w_m\|^2 &= (w_m, w_m) = (w_m, w_m - w_n) + (w_m, w_n) \\
&= (w_m, w_m - w_n) + H_n(w_m^*)^* \\
&\leq \varepsilon \|w_m\| + |H_n(w_m^*)^*|.
\end{aligned}$$

Letting $n \to \infty$, from (15) we get

$$\|w_m\|^2 \leq \varepsilon \|w_m\|,$$

or

$$\|w_m\| \leq \varepsilon, \qquad m \geq N. \qquad \square$$

We *define* $\mathbf{L}^2$ to be the set consisting of all periodic distributions F with the property that there is a sequence $(u_n)_1^\infty \subset \mathscr{C}$ such that

$$\begin{aligned}
\|u_n - u_m\| &\to 0 \quad \text{as} \quad n, m \to \infty, \\
F_{u_n} &\to F \; (\mathscr{P}').
\end{aligned}$$

If $(u_n)_1^\infty \subset \mathscr{C}$ is such a sequence, we say that it *converges to F in the sense of* $\mathbf{L}^2$ and write

$$u_n \to F \; (\mathbf{L}^2).$$

Lemma 1.2 can be rephrased: if

$$u_n \to F \; (\mathbf{L}^2), \qquad v_n \to G \; (\mathbf{L}^2),$$

then

$$F = G \qquad \text{if and only if } \|u_n - v_n\| \to 0.$$

Clearly $\mathbf{L}^2$ is a subspace of $\mathscr{P}'$ in the sense of vector spaces. In fact, if

$$u_n \to F \; (\mathbf{L}^2) \quad \text{and} \quad v_n \to G \; (\mathbf{L}^2),$$

then

$$au_n \to aF \ (L^2), \qquad u_n + v_n \to F + G \ (L^2).$$

We may extend the inner product on $\mathscr{C}$ to L^2 as follows. If

$$u_n \to F \ (L^2), \qquad v_n \to G \ (L^2),$$

let

(17) $$(F, G) = \lim (u_n, v_n).$$

The existence of this limit is left as an exercise. Lemma 1.2 shows that the limit is independent of the particular sequences $(u_n)_1^\infty$ and $(v_n)_1^\infty$. That is, if also

$$u_n' \to F \ (L^2), \qquad v_n' \to G \ (L^2)$$

then

(18) $$\lim (u_n, v_n) = \lim (u_n', v_n').$$

Theorem 1.3. *The inner product in L^2 defined by (17) satisfies the identities (2), (3), (4), (5), (6). If we define*

(19) $$\|F\| = (F, F)^{1/2},$$

then this is a norm on L^2. The space L^2 is complete with respect to this norm.

Proof. The fact that (2)–(6) hold is a consequence of (17) and (2)–(6) for functions. We also have the Schwarz inequality in L^2:

$$|(F, G)| \le \|F\| \, \|G\|.$$

It follows that $\|F\|$ is a norm.

Finally, suppose $(F_n)_1^\infty \subset L^2$ is a Cauchy sequence with respect to this norm. First, note that if

$$u_n \to F \ (L^2)$$

and $v \in \mathscr{C}$, and if we take $v_n = v$, all n, then

$$\|F_v - F\|^2 = \lim \|v_n - u_n\|^2 = \lim \|v - u_n\|^2.$$

It follows that

$$\|F_{u_n} - F\|^2 \to 0 \quad \text{as} \quad n \to \infty,$$

i.e., F can be approximated in L^2 by functions. Therefore, for each $n = 1, 2, \ldots$ we can find a function $v_n \in \mathscr{C}$ such that

$$\|F_n - F_{v_n}\| < n^{-1}.$$

Then

$$\|v_n - v_m\| = \|F_{v_n} - F_{v_m}\| \le \|F_{v_n} - F_n\| + \|F_n - F_m\| + \|F_m - F_{v_m}\|$$
$$< \|F_n - F_m\| + n^{-1} + m^{-1} \to 0 \quad \text{as} \quad n, m \to \infty.$$

Thus there is an $F \in \mathbf{L}^2$ such that

$$v_n \to F \quad (\mathbf{L}^2).$$

But then

$$\|F_n - F\| \le \|F_n - F_{v_n}\| + \|F_{v_n} - F\|$$
$$< n^{-1} + \|F_{v_n} - F\| \to 0. \qquad \Box$$

Exercises

1. Carry out the proof that $\mathscr{C}$ is not complete with respect to the norm $\|u\|$.

2. Show that the limit in (17) exists.

3. Show that (18) is true.

4. Suppose $f: [0, 2\pi] \to \mathbb{R}$ is such that

$$f(x) = 1, \qquad x \in [a, b),$$
$$f(x) = 0, \qquad x \notin [a, b).$$

Define

$$F(v) = \frac{1}{2\pi} \int_0^{2\pi} f(x)v(x)\,dx = \frac{1}{2\pi} \int_a^b v(x)\,dx, \qquad v \in \mathscr{P}.$$

Show that $F \in \mathbf{L}^2$.

5. Suppose $f: [0, 2\pi] \to \mathbb{C}$ is constant on each subinterval $[x_{i-1}, x_i)$, where

$$0 = x_0 < x_1 < \cdots < x_n = 2\pi.$$

Define

$$F(v) = \frac{1}{2\pi} \int_0^{2\pi} f(x)v(x)\,dx.$$

Show that $F \in \mathbf{L}^2$.

6. Show that δ, the δ-distribution, is *not* in $\mathbf{L}^2$.

7. Show that if $F \in \mathbf{L}^2$ there is a sequence $(u_n)_1^\infty$ of *smooth* periodic functions such that

$$u_n \to F \quad (\mathbf{L}^2).$$

8. Let T_t denote translation. Show that if $u \in \mathscr{C}$ then

$$\|T_t u - T_s u\| \to 0 \quad \text{as} \quad t \to s.$$

If $F \in \mathbf{L}^2$, show that

$$\|T_t F - T_s F\| \to 0 \quad \text{as} \quad t \to s.$$

9. For any $F \in \mathbf{L}^2$, show that

$$\|F\| = \sup \{|F(u)| \mid u \in \mathscr{P}, \|u\| \le 1\}.$$

§2. Hilbert space

In this section we consider an abstract version of the space L^2 of §1. This clarifies the nature of certain theorems. In addition, the abstract version describes other spaces which are obtained in very different ways.

Suppose H is a vector space over the real or complex numbers. An *inner product* in H is a function assigning to each ordered pair of elements $\mathbf{u}, \mathbf{v} \in H$ a real or complex number denoted by $(\mathbf{u}, \mathbf{v})$, such that

$$(a\mathbf{u}, \mathbf{v}) = a(\mathbf{u}, \mathbf{v}), \quad a \text{ a scalar,}$$
$$(\mathbf{u}_1 + \mathbf{u}_2, \mathbf{v}) = (\mathbf{u}_1, \mathbf{v}) + (\mathbf{u}_2, \mathbf{v}),$$
$$(\mathbf{v}, \mathbf{u}) = (\mathbf{u}, \mathbf{v})^* \quad ((\mathbf{v}, \mathbf{u}) = (\mathbf{u}, \mathbf{v}) \text{ in the real case}),$$
$$(\mathbf{u}, \mathbf{u}) > 0 \quad \text{if } \mathbf{u} \neq 0.$$

The argument of Lemma 1.1 shows that

(1)
$$|(\mathbf{u}, \mathbf{v})| \leq \|\mathbf{u}\|\, \|\mathbf{v}\|,$$

where

(2)
$$\|\mathbf{u}\| = (\mathbf{u}, \mathbf{u})^{1/2}.$$

Then $\|\mathbf{u}\|$ is a norm on H. If H is complete with respect to the metric associated with this norm, then H is said to be a *Hilbert space*. In particular, L^2 is a Hilbert space. Clearly any Hilbert space is a Banach space.

A more mundane example than L^2 is the finite-dimensional vector space $\mathbb{C}^N$ of N-tuples of complex numbers, with

$$(\mathbf{a}, \mathbf{b}) = \sum_{n=1}^{N} a_n b_n^*$$

when

$$\mathbf{a} = (a_1, a_2, \ldots, a_N), \quad \mathbf{b} = (b_1, b_2, \ldots, b_N).$$

In this case the Schwarz inequality is

$$\left| \sum_{n=1}^{N} a_n b_n^* \right|^2 \leq \sum_{n=1}^{N} |a_n|^2 \sum_{n=1}^{N} |b_n|^2.$$

Notice in particular that if we let

$$\mathbf{a}' = (|a_1|, |a_2|, \ldots, |a_N|), \quad \mathbf{b}' = (|b_1|, |b_2|, \ldots, |b_N|),$$

then

$$\left(\sum_{n=1}^{N} |a_n|\, |b_n| \right)^2 \leq \sum_{n=1}^{N} |a_n|^2 \sum_{n=1}^{N} |b_n|^2.$$

A still more mundane example is the plane $\mathbb{R}^2$, with

$$\mathbf{a} \cdot \mathbf{b} = a_1 b_1 + a_2 b_2$$

when $\mathbf{a} = (a_1, a_2)$, $\mathbf{b} = (b_1, b_2)$; here we use the dot to avoid confusing the inner product with ordered pairs. It is worth noting that the *law of cosines* of trigonometry can be written

$$\mathbf{a} \cdot \mathbf{b} = |\mathbf{a}|\, |\mathbf{b}| \cos \theta,$$

where θ is the angle between the line segments from $\mathbf{0}$ to $\mathbf{a}$ and the line segment from $\mathbf{0}$ to $\mathbf{b}$. Therefore $|\mathbf{a} \cdot \mathbf{b}| = |\mathbf{a}| \, |\mathbf{b}|$ if and only if the segments lie on the same line. Similarly, $\mathbf{a} \cdot \mathbf{b} = 0$ if and only if the segments form a right angle.

Elements $\mathbf{u}$ and $\mathbf{v}$ of a Hilbert space $\mathbf{H}$ are said to be *orthogonal* if the inner product $(\mathbf{u}, \mathbf{v})$ is zero. In $\mathbb{R}^2$ this means that the corresponding line segments are perpendicular. We write

$$\mathbf{u} \perp \mathbf{v}$$

when $\mathbf{u}$ and $\mathbf{v}$ are orthogonal. More generally, $\mathbf{u} \in \mathbf{H}$ is said to be orthogonal to the subset $S \subset \mathbf{H}$ if

$$\mathbf{u} \perp \mathbf{v}, \qquad \text{all } \mathbf{v} \in S.$$

If so we write

$$\mathbf{u} \perp S.$$

If $\mathbf{u} \perp \mathbf{v}$ then

$$(3) \qquad \begin{aligned} \|\mathbf{u} + \mathbf{v}\|^2 &= (\mathbf{u} + \mathbf{v}, \mathbf{u} + \mathbf{v}) = (\mathbf{u}, \mathbf{u}) + (\mathbf{u}, \mathbf{v}) + (\mathbf{v}, \mathbf{u}) + (\mathbf{v}, \mathbf{v}) \\ &= \|\mathbf{u}\|^2 + \|\mathbf{v}\|^2. \end{aligned}$$

In $\mathbb{R}^2$, this is essentially the *Pythagorean theorem*, and we shall give the identity (3) that name in any case. Another simple identity with a classical geometric interpretation is the *parallelogram law*:

$$(4) \qquad \|\mathbf{u} - \mathbf{v}\|^2 + \|\mathbf{u} + \mathbf{v}\|^2 = 2\|\mathbf{u}\|^2 + 2\|\mathbf{v}\|^2.$$

This follows immediately from the properties of the inner product. In $\mathbb{R}^2$ it says that the sum of the squares of the lengths of the diagonals of the parallelogram with vertices $\mathbf{0}, \mathbf{u}, \mathbf{v}, \mathbf{u} + \mathbf{v}$ is equal to the sum of the lengths of the squares of the (four) sides.

When speaking of *convergence* in a Hilbert space, we shall always mean convergence with respect to the metric associated with the norm. Thus

$$\mathbf{u}_n \to \mathbf{u} \quad \text{means} \quad \|\mathbf{u}_n - \mathbf{u}\| \to 0.$$

The Schwarz inequality shows that the inner product is a continuous function.

Lemma 2.1. *If* $(\mathbf{u}_n)_1^\infty, (\mathbf{v}_n)_1^\infty \subset \mathbf{H}$ *and*

$$\mathbf{u}_n \to \mathbf{u}, \qquad \mathbf{v}_n \to \mathbf{v},$$

then

$$(\mathbf{u}_n, \mathbf{v}_n) \to (\mathbf{u}, \mathbf{v}).$$

Proof. Since the sequences converge, they are bounded. In particular, there is a constant M such that $\|\mathbf{v}_n\| \le M$, all n. Then

$$\begin{aligned} |(\mathbf{u}_n, \mathbf{v}_n) - (\mathbf{u}, \mathbf{v})| &= |(\mathbf{u}_n - \mathbf{u}, \mathbf{v}_n) + (\mathbf{u}, \mathbf{v}_n - \mathbf{v})| \\ &\le \|\mathbf{u}_n - \mathbf{u}\| \, \|\mathbf{v}_n\| + \|\mathbf{u}\| \, \|\mathbf{v}_n - \mathbf{v}\| \\ &\le M\|\mathbf{u}_n - \mathbf{u}\| + \|\mathbf{u}\| \, \|\mathbf{v}_n - \mathbf{v}\| \to 0. \qquad \square \end{aligned}$$

Corollary 2.2. *If* $\mathbf{u} \in \mathbf{H}$, $S \subset \mathbf{H}$, *and* $\mathbf{u} \perp S$, *then* $\mathbf{u}$ *is orthogonal to the closure of* S.

The general theory of Hilbert space essentially rests on the following two geometric lemmas.

Lemma 2.3. *Suppose* $\mathbf{H}_1$ *is a closed subspace of the Hilbert space* $\mathbf{H}$, *and suppose* $\mathbf{u} \in \mathbf{H}$. *Then there is a unique* $\mathbf{v} \in \mathbf{H}_1$ *which is closest to* $\mathbf{u}$, *in the sense that*

$$\|\mathbf{u} - \mathbf{v}\| \le \|\mathbf{u} - \mathbf{w}\|, \qquad \text{all } \mathbf{w} \in \mathbf{H}_1.$$

Proof. The set

$$\{\|\mathbf{u} - \mathbf{w}\| \mid \mathbf{w} \in \mathbf{H}_1\}$$

is bounded below by 0. Let d be the greatest lower bound of this set. For each integer $n > 0$ there is an element $\mathbf{v}_n \in \mathbf{H}_1$ such that

$$\|\mathbf{u} - \mathbf{v}_n\| < d + n^{-1}.$$

If we can show that $(\mathbf{v}_n)_1^\infty$ is a Cauchy sequence, then it has a limit $\mathbf{v} \in \mathbf{H}$. Since $\mathbf{H}_1$ is closed, we would have $\mathbf{v} \in \mathbf{H}_1$ and $\|\mathbf{u} - \mathbf{v}\| = d$ as desired.

Geometrically the argument that $(\mathbf{v}_n)_1^\infty$ is a Cauchy sequence is as follows. The midpoint $\frac{1}{2}(\mathbf{v}_n + \mathbf{v}_m)$ of the line segment joining $\mathbf{v}_n$ and $\mathbf{v}_m$ has distance $\ge d$ from $\mathbf{u}$, by the definition of d. Therefore the square of the length of one diagonal of the parallelogram with vertices $\mathbf{u}$, $\mathbf{v}_n$, $\mathbf{v}_m$, $\mathbf{v}_n + \mathbf{v}_m$ is nearly equal to the sum of squares of the lengths of the sides. It follows that the length $\|\mathbf{v}_n - \mathbf{v}_m\|$ of the other diagonal is small. Algebraically, we use (4) to get

$$\begin{aligned} 0 \le \|\mathbf{v}_n - \mathbf{v}_m\|^2 &= 2\|\mathbf{v}_n - \mathbf{u}\|^2 + 2\|\mathbf{v}_m - \mathbf{u}\|^2 - \|(\mathbf{v}_n + \mathbf{v}_m) - 2\mathbf{u}\|^2 \\ &< 2(d + n^{-1})^2 + 2(d + m^{-1})^2 - 4\|\tfrac{1}{2}(\mathbf{v}_n + \mathbf{v}_m) - \mathbf{u}\|^2 \\ &\le 2(d + n^{-1})^2 + 2(d + m^{-1})^2 - 4d^2 \to 0. \end{aligned}$$

To show uniqueness, suppose that $\mathbf{v}$ and $\mathbf{w}$ both are closest to $\mathbf{u}$ in the above sense. Then another application of the parallelogram law gives

$$\begin{aligned} \|\mathbf{u} - \tfrac{1}{2}(\mathbf{v} + \mathbf{w})\|^2 &= \tfrac{1}{2}\|\mathbf{u} - \mathbf{v}\|^2 + \tfrac{1}{2}\|\mathbf{u} - \mathbf{w}\|^2 - \tfrac{1}{4}\|\mathbf{v} - \mathbf{w}\|^2 \\ &= d^2 - \|\mathbf{v} - \mathbf{w}\|^2. \end{aligned}$$

Since the left side is $\ge d^2$, we must have $\mathbf{v} = \mathbf{w}$. $\square$

As an example, take $\mathbf{H} = \mathbb{R}^2$, $\mathbf{H}_1$ a line through the origin. The unique point on this line closest to a given point $\mathbf{u}$ is obtained as the intersection of $\mathbf{H}_1$ and the line through $\mathbf{u}$ perpendicular to $\mathbf{H}_1$. This connection between perpendicularity (orthogonality) and the closest point is also true in the general case.

Lemma 2.4. *Under the hypotheses of Lemma 2.3, the element* $\mathbf{v} \in \mathbf{H}_1$ *is closest to* $\mathbf{u}$ *if and only if*

$$\mathbf{u} - \mathbf{v} \perp \mathbf{H}_1.$$

Proof. First, suppose $v \in H_1$ is closest to u, and suppose $w \in H_1$. We want to show $(u - v, w) = 0$, and we may assume $w \neq 0$. Let $u_1 = u - v$. For any $a \in \mathbb{C}$, $v + aw \in H_1$. Therefore

$$\|u_1 - aw\|^2 = \|u - (v + aw)\|^2 \geq \|u - v\|^2 = \|u_1\|^2,$$

or

$$\|u_1\|^2 - (u_1, aw) - (aw, u_1) + |a|^2 \|w\|^2 \geq \|u_1\|^2.$$

Let

(5) $$a = (u_1, w)\|w\|^{-2}.$$

Then (5) becomes

$$-|(u_1, w)|^2 \|w\|^{-2} \geq 0.$$

Thus $(u_1, w) = 0$.

Conversely, suppose $u - v \perp H_1$, and suppose $w \in H_1$. Then $v - w \in H_1$, so

$$(u - v) \perp (v - w).$$

The Pythagorean theorem gives

$$\|u - w\|^2 = \|(u - v) + (v - w)\|^2 = \|u - v\|^2 + \|v - w\|^2$$
$$\geq \|u - v\|^2. \qquad \square$$

Corollary 2.5. *Suppose* H_1 *is a closed subspace of a Hilbert space* H. *Then either* $H_1 = H$, *or there is a nonzero element* $u \in H$ *such that* $u \perp H_1$.

Proof. If $H_1 \neq H$, take $u_0 \in H$, $u_0 \notin H_1$. Take $v_0 \in H_1$ such that v_0 is closest to u_0. Then $u = u_0 - v_0$ is nonzero and orthogonal to H_1. $\square$

As a first application of these results, we determine all the bounded linear functionals on H. The following theorem is one of several results known as the *Riesz Representation Theorem*.

Theorem 2.6. *Suppose* H *is a Hilbert space and suppose* $v \in H$. *The mapping* $L_v : H \to \mathbb{C}$ *(or* $\mathbb{R}$*) defined by*

$$L_v(u) = (u, v), \qquad u \in H,$$

is a bounded linear functional on H. *Moreover, if* L *is any bounded linear functional on* H, *then there is a unique* $v \in H$ *such that* $L = L_v$.

Proof. Clearly L_v is linear. By the Schwarz inequality

$$|L_v(u)| \leq \|v\| \|u\|.$$

Thus L_v is bounded.

Suppose L is a bounded linear functional on H. If $L = 0$ we may take $v = 0$. Otherwise, let

$$H_1 = \{u \in H \mid L(u) = 0\}.$$

Then H_1 is a subspace of H, since L is linear; H_1 is closed, since L is continuous. Since $L \neq 0$, H_1 is not H. Take a nonzero $u \in H$ which is orthogonal to H_1, and let

$$v = \|u\|^{-2}L(u)^*u.$$

Then also $v \perp H_1$, so

$$L_v(w) = L(w), \qquad w \in H_1.$$

Moreover,

$$L_v(u) = \|u\|^{-2}L(u)(u, u) = L(u).$$

If w is any element of H,

$$w - L(u)^{-1}L(w)u \in H_1.$$

Thus any element of H is of the form

$$au + w_1$$

for some $a \in \mathbb{C}$ (or $\mathbb{R}$) and $w_1 \in H_1$. It follows that $L_v = L$.

To show uniqueness, suppose $L_v = L_w$. Then

$$0 = L_v(v - w) - L_w(v - w) = \|v - w\|^2. \qquad \square$$

Exercises

1. Prove the law of cosines as stated above.

2. Suppose $F \in \mathscr{P}'$. Show that $F \in L^2$ if and only if there is a constant c such that

$$|F(u)| \le c\|u\|, \qquad \text{all } u \in \mathscr{P}.$$

3. Let H be any Hilbert space and let H_1 be a closed subspace of H. Let

$$H_2 = \{u \in H \mid u \perp H_1\}.$$

Show that H_2 is a closed subspace of H. Show that for any $u \in H$ there are unique vectors $u_1 \in H_1$ and $u_2 \in H_2$ such that

$$u = u_1 + u_2.$$

§3. Hilbert spaces of sequences

In this section we consider two infinite dimensional analogs of the finite dimensional complex Hilbert space $\mathbb{C}^N$. Recall that if

$$x = (a_1, a_2, \ldots, a_N) \in \mathbb{C}^N$$

then

$$\|x\|^2 = \sum_{n=1}^{N} |a_n|^2.$$

Let $l_+{}^2$ denote the set of all sequences

$$\mathbf{x} = (a_n)_1^\infty \subset \mathbb{C}$$

such that

(1)
$$\sum_{n=1}^\infty |a_n|^2 < \infty.$$

If

$$\mathbf{x} = (a_n)_1^\infty \in l_+{}^2, \qquad \mathbf{y} = (b_n)_1^\infty \in l_+{}^2,$$

we set

(2)
$$(\mathbf{x}, \mathbf{y}) = \sum_{n=1}^\infty a_n b_n^*,$$

provided this series converges.

Theorem 3.1. *The space $l_+{}^2$ of complex sequences satisfying* (1) *is a Hilbert space with respect to the inner product* (2).

Proof. Suppose

$$\mathbf{x} = (a_n)_1^\infty \in l_+{}^2 \quad \text{and} \quad \mathbf{y} = (b_n)_1^\infty \in l_+{}^2.$$

If $a \in \mathbb{C}$, then clearly

$$a\mathbf{x} = (aa_n)_1^\infty \in l_+{}^2.$$

As for $\mathbf{x} + \mathbf{y}$, we have

$$\sum |a_n + b_n|^2 \le \sum (|a_n|^2 + 2|a_n b_n| + |b_n|^2)$$
$$\le 2 \sum |a_n|^2 + 2 \sum |b_n|^2 < \infty.$$

Thus $l_+{}^2$ is a vector space. To show that the inner product (2) is defined for all $\mathbf{x}, \mathbf{y} \in l_+{}^2$, we use the inequality (1) from §2. For each N,

$$\sum_{n=1}^N |a_n b_n^*| \le \left(\sum_{n=1}^N |a_n|^2 \right)^{1/2} \left(\sum_{n=1}^N |b_n|^2 \right)^{1/2}$$
$$\le \left(\sum_{n=1}^\infty |a_n|^2 \right)^{1/2} \left(\sum_{n=1}^\infty |b_n|^2 \right)^{1/2}$$

Therefore

$$\sum_{n=1}^\infty |a_n b_n^*| < \infty$$

and (2) converges. It is easy to check that $(\mathbf{x}, \mathbf{y})$ has the properties of an inner product. The only remaining question is whether $l_+{}^2$ is complete.

Suppose

$$\mathbf{x}_m = (a_{m,n})_{n=1}^\infty \in l_+{}^2, \qquad m = 1, 2, \ldots .$$

Suppose $(x_m)_1^\infty$ is a Cauchy sequence in the metric corresponding to the norm

$$\|x\| = (x, x)^{1/2}.$$

For each fixed n,

$$|a_{m,n} - a_{p,n}|^2 \leq \|x_m - x_p\|^2 \to 0$$

as $m, p \to \infty$. Thus $(a_{m,n})_{m=1}^\infty$ is a Cauchy sequence in $\mathbb{C}$, and

$$a_{m,n} \to a_n \quad \text{as} \quad m \to \infty.$$

Let $x = (a_n)_1^\infty$. We want to show that $x \in l_+^2$ and $\|x_m - x\| \to 0$. Since $(x_m)_1^\infty$ is a Cauchy sequence, it is bounded:

$$\|x_m\| \leq K, \quad \text{all } m.$$

Therefore for any N,

$$\sum_{n=1}^N |a_n|^2 = \lim_{m \to \infty} \sum_{n=1}^N |a_{m,n}|^2 \leq K^2.$$

Finally, given $\varepsilon > 0$ choose M so large that $m, p \geq M$ implies

$$\|x_m - x_p\| < \varepsilon.$$

Then for any N and any $m \geq M$,

$$\sum_{n=1}^N |a_{m,n} - a_n|^2 = \lim_{p \to \infty} \sum_{n=1}^N |a_{m,n} - a_{p,n}|^2 \leq \varepsilon^2.$$

Thus

$$\|x_m - x\| \leq \varepsilon \quad \text{if } m \geq M. \qquad \square$$

It is often convenient to work with sequences indexed by the integers, rather than by the positive integers; such sequences are called *two-sided sequences*. We use the notation

$$x = (a_n)_{-\infty}^\infty = (\ldots, a_{-2}, a_{-1}, a_0, a_1, a_2, \ldots).$$

Let l^2 denote the space of two-sided sequences

$$x = (a_n)_{-\infty}^\infty \subset \mathbb{C}$$

such that

(3)
$$\sum_{n=-\infty}^\infty |a_n|^2 < \infty.$$

Here a two-sided infinite sum is defined to be the limit

$$\sum_{n=-\infty}^\infty c_n = \lim_{M,N \to +\infty} \sum_{n=-M}^N c_n$$

if this limit exists.
If

$$x = (a_n)_{-\infty}^\infty \in l^2, \qquad y = (b_n)_{-\infty}^\infty \in l^2,$$

we let

(4)
$$(\mathbf{x}, \mathbf{y}) = \sum_{n=-\infty}^{\infty} a_n b_n^*,$$

provided the series converges.

Theorem 3.2. *The space l^2 of two-sided complex sequences satisfying (3) is a Hilbert space with respect to the inner product (4).*

The proof of this theorem is very similar to the proof of Theorem 3.1.

Exercises

1. Prove Theorem 3.2
2. Let $\mathbf{e}_m \in l_+{}^2$ be the sequence

$$\mathbf{e}_m = (a_{m,n})_{n=1}^{\infty} .$$

with $a_{m,n} = 0$ if $m \neq n$, $a_{n,n} = 1$. Show that

(a) $\|e_m\| = 1$;
(b) $\mathbf{e}_m \perp \mathbf{e}_p$ if $m \neq p$;
(c) $(x, \mathbf{e}_m) \to 0$ as $m \to \infty$, for each $x \in l_+{}^2$;
(d) the set of linear combinations of the elements $\mathbf{e}_m$ is dense in $l_+{}^2$;
(e) if $\mathbf{x} \in l_+{}^2$ there is a unique sequence $(b_n)_1^{\infty} \subset \mathbb{C}$ such that

$$\left\| \mathbf{x} - \sum_{n=1}^{N} b_n \mathbf{e}_n \right\| \to 0 \quad \text{as} \quad N \to \infty.$$

3. Show that the unit ball in $l_+{}^2$, the set

$$B = \{ \mathbf{x} \in l_+{}^2 \mid \|\mathbf{x}\| \leq 1 \},$$

is closed and bounded, but not compact.
4. Show that the set

$$C = \{ \mathbf{x} \in l_+{}^2 \mid \mathbf{x} = (a_n)_1^{\infty}, \text{ each } |a_n| \leq n^{-1} \}$$

is compact; C is called the *Hilbert cube*.

§4. Orthonormal bases

The Hilbert space $\mathbb{C}^N$ is a finite dimensional vector space. Therefore any element of $\mathbb{C}^N$ can be written uniquely as a linear combination of a given set of basis vectors. It follows that the inner product of two elements of $\mathbb{C}^N$ can be computed if we know the expression of each element as such a linear combination. Conversely, the inner product makes possible a very convenient way of expressing a given vector as a linear combination of basis vectors.

Specifically, let $\mathbf{e}_n \in \mathbb{C}^N$ be the N-tuple

$$\mathbf{e}_n = (0, 0, \ldots, 0, 1, 0, \ldots, 0),$$

where the 1 is in the n-th place. Then $\{\mathbf{e}_1, \mathbf{e}_2, \ldots, \mathbf{e}_N\}$ is a basis for $\mathbb{C}^N$. Moreover it is clear that

(1) $(\mathbf{e}_n, \mathbf{e}_m) = 1$ if $n = m$, $(\mathbf{e}_n, \mathbf{e}_m) = 0$ if $n \neq m$.

If $\mathbf{x} = (a_1, a_2, \ldots, a_N) \in \mathbb{C}^N$ then the expression for $\mathbf{x}$ as a linear combination of the basis vectors $\mathbf{e}_n$ is

$$(2) \qquad \mathbf{x} = \sum_{n=1}^{N} a_n \mathbf{e}_n.$$

Because of (1),

$$(\mathbf{x}, \mathbf{e}_m) = a_m.$$

Thus we may rewrite (2) as

$$(3) \qquad \mathbf{x} = \sum_{n=1}^{N} (\mathbf{x}, \mathbf{e}_n)\mathbf{e}_n.$$

If $\mathbf{x} = (a_1, a_2, \ldots, a_N)$ and $\mathbf{y} = (b_1, b_2, \ldots, b_N)$, then

$$(\mathbf{x}, \mathbf{y}) = \sum_{n=1}^{N} a_n b_n^*.$$

Using (3) and the corresponding expression for y, we have

$$(4) \qquad (\mathbf{x}, \mathbf{y}) = \sum_{n=1}^{N} (\mathbf{x}, \mathbf{e}_n)(\mathbf{y}, \mathbf{e}_n)^* = \sum_{n=1}^{N} (\mathbf{x}, \mathbf{e}_n)(\mathbf{e}_n, \mathbf{y}).$$

In particular,

$$(5) \qquad \|\mathbf{x}\|^2 = \sum_{n=1}^{N} |(\mathbf{x}, \mathbf{e}_n)|^2.$$

The aim of this section and the next is to carry this development over to a class of Hilbert spaces which are not finite dimensional. We look for infinite subsets $(\mathbf{e}_n)_1^\infty$ with the properties (1), and try to write elements as convergent infinite sums analogous to (3).

A subset S of a Hilbert space $\mathbf{H}$ is said to be *orthonormal* if each $\mathbf{u} \in S$ has norm 1, while

$$(\mathbf{u}, \mathbf{v}) = 0 \qquad \text{if } \mathbf{u}, \mathbf{v} \in S, \mathbf{u} \neq \mathbf{v}.$$

The following procedure for producing orthonormal sets is called the *Gram-Schmidt method*.

Lemma 4.1. *Suppose $\{\mathbf{u}_1, \mathbf{u}_2, \ldots\}$ is a finite or countable set of elements of a Hilbert space $\mathbf{H}$. Then there is a finite or countable set $S = \{\mathbf{e}_1, \mathbf{e}_2, \ldots\}$ of elements of $\mathbf{H}$ such that S is orthonormal and such that each $\mathbf{u}_n$ is in the subspace spanned by $\{\mathbf{e}_1, \mathbf{e}_2, \ldots, \mathbf{e}_n\}$.*

(If S has m elements, $m < \infty$, we interpret the statement as saying that $\mathbf{u}_n \in \text{span}\{\mathbf{e}_1, \ldots, \mathbf{e}_m\}$ when $n \geq m$.)

Proof. The proof is by induction. If each $\mathbf{u}_i = 0$, we may take S to be the empty set. Otherwise let $\mathbf{v}_1$ be the first nonzero $\mathbf{u}_i$, and let

$$\mathbf{e}_1 = \|\mathbf{v}_1\|^{-1}\mathbf{v}_1.$$

Then $\{\mathbf{e}_1\}$ is orthonormal and $\mathbf{u}_1 \in \text{span}\{\mathbf{e}_1\}$. Suppose we have chosen $\mathbf{e}_1, \ldots, \mathbf{e}_m$ such that $\{\mathbf{e}_1, \ldots, \mathbf{e}_m\}$ is orthonormal and $\mathbf{u}_1, \ldots, \mathbf{u}_m \in \text{span}\{\mathbf{e}_1, \ldots, \mathbf{e}_m\}$. If each $\mathbf{u}_i \in \text{span}\{\mathbf{e}_1, \ldots, \mathbf{e}_m\}$ we may stop. Otherwise choose the first j such that $\mathbf{u}_j$ is not in this subspace. Let

$$\mathbf{v}_{m+1} = \mathbf{u}_j - \sum_{n=1}^{m} (\mathbf{u}_j, \mathbf{e}_n)\mathbf{e}_n.$$

Since $\{\mathbf{e}_1, \ldots, \mathbf{e}_m\}$ is orthonormal, it follows that

$$(\mathbf{v}_{m+1}, \mathbf{e}_n) = 0, \qquad 1 \leq n \leq m.$$

Since $\mathbf{u}_j \notin \text{span}\{\mathbf{e}_1, \ldots, \mathbf{e}_m\}$, $\mathbf{v}_{m+1} \neq 0$. Let

$$\mathbf{e}_{m+1} = \|\mathbf{v}_{m+1}\|^{-1}\mathbf{v}_{m+1}.$$

Then $\{\mathbf{e}_1, \ldots, \mathbf{e}_{m+1}\}$ is orthonormal and $\mathbf{u}_{m+1}$ is in the span. Continuing, we get the desired set S. □

Note that completeness of $\mathbf{H}$ was not used. Thus Lemma 4.1 is valid in any space with an inner product.

An *orthonormal basis* for a Hilbert space $\mathbf{H}$ is an orthonormal set $S \subset \mathbf{H}$ such that span (S) is dense in $\mathbf{H}$. This means that for any $\mathbf{u} \in \mathbf{H}$ and any $\varepsilon > 0$, there is a $\mathbf{v}$, which is a linear combination of elements of S, such that $\|\mathbf{u} - \mathbf{v}\| < \varepsilon$.

A Hilbert space $\mathbf{H}$ is said to be *separable* if there is a sequence $(\mathbf{u}_n)_1^\infty \subset \mathbf{H}$ which is dense in $\mathbf{H}$. This means that for any $\mathbf{u} \in \mathbf{H}$ and any $\varepsilon > 0$, there is an n such that $\|\mathbf{u} - \mathbf{u}_n\| < \varepsilon$.

Theorem 4.2. *Suppose $\mathbf{H}$ is a separable Hilbert space. Then $\mathbf{H}$ has an orthonormal basis S, which is finite or countable.*

Conversely, if $\mathbf{H}$ is a Hilbert space which has a finite or countable orthonormal basis, then $\mathbf{H}$ is separable.

Proof. Suppose $(\mathbf{u}_n)_1^\infty$ is dense in $\mathbf{H}$. By Lemma 4.1, there is a finite or countable orthonormal set $S = \{\mathbf{e}_1, \mathbf{e}_2, \ldots\}$ such that each $\mathbf{u}_n$ is a linear combination of elements of S. Thus S is an orthonormal basis.

Conversely, suppose S is a finite or countable orthonormal basis for $\mathbf{H}$. Suppose $\mathbf{H}$ is a complex vector space. Let T be the set of all elements of $\mathbf{H}$ of the form

$$\mathbf{u} = \sum_{n=1}^{N} a_n\mathbf{e}_n,$$

where N is arbitrary, the e_n are in S, and the a_n are complex numbers whose real and imaginary parts are *rational*. It is not difficult to show that T is countable, so the elements of T may be arranged in a sequence $(u_n)_1^\infty$. Any complex number is the limit of a sequence of complex numbers with rational real and imaginary parts. It follows that any linear combination of elements of S is a limit of a sequence of elements in T. Since S is assumed to be an orthonormal basis, this implies that $(u_n)_1^\infty$ is dense in H. □

To complete Theorem 4.2, we want to know whether two orthonormal bases in a separable Hilbert space have the same number of elements.

Theorem 4.3. *Suppose H is a separable Hilbert space. If $\dim H = N < \infty$, then any orthonormal basis for H is a basis for H as a vector space, and therefore has N elements.*

If H is not finite dimensional, then any orthonormal basis for H is countable.

Proof. Suppose $\dim H = N < \infty$, and suppose $S \subseteq H$ is an orthonormal basis. If $e_1, \ldots, e_M$ are distinct elements of S and

$$\sum_{n=1}^{N} a_n e_n = 0,$$

then

$$0 = \left(\sum a_n e_n, e_m \right) = a_m, \qquad m = 1, \ldots, M.$$

Thus the elements of S are linearly independent, so S has $\leq N$ elements. Let $S = \{e_1, e_2, \ldots, e_M\}$. We want to show that S is a basis. Let H_1 be the subspace spanned by S. Given $u \in H$, let

$$u_1 = \sum_{n=1}^{M} (u, e_n) e_n.$$

Then

$$(u - u_1, e_m) = 0, \qquad m = 1, \ldots, M.$$

It follows that $u - u_1$ is orthogonal to the subspace H_1. The argument used to prove Lemma 2.4 shows that u_1 is the element of H_1 closest to u. But by assumption on S, there are elements of H_1 arbitrarily close to u. Therefore $u = u_1 \in H_1$, and S is a basis.

The argument just given shows that if H has a finite orthonormal basis S, then S is a basis in the vector space sense. Therefore if H is not finite dimensional, any orthonormal basis is infinite. We want to show, therefore, that if H is separable and not finite dimensional then any orthonormal basis is at most countable. Let $(u_n)_1^\infty$ be dense in H, and let S be an orthonormal basis. For each element $e \in S$, there is an integer $n = n(e)$ such that

$$\|e - u_n\| < 2^{-1/2}.$$

Suppose $e, f \in S$ and $e \neq f$. Let $n = n(e)$, $p = n(f)$. Then

$$\|e - f\|^2 = \|e\|^2 + (e, f) + (f, e) + \|f\|^2$$
$$= 1 + 0 + 0 + 1 = 2,$$

so

$$\begin{aligned}
\|\mathbf{u}_n - \mathbf{u}_p\| &= \|\mathbf{u}_n - \mathbf{e} + \mathbf{e} - \mathbf{f} + \mathbf{f} - \mathbf{u}_p\| \\
&\geq \|\mathbf{e} - \mathbf{f}\| - \|\mathbf{u}_n - \mathbf{e} + \mathbf{f} - \mathbf{u}_p\| \\
&\geq \|\mathbf{e} - \mathbf{f}\| - \|\mathbf{u}_n - \mathbf{e}\| - \|\mathbf{f} - \mathbf{u}_p\| \\
&> 2^{1/2} - 2^{-1/2} - 2^{-1/2} = 0
\end{aligned}$$

Thus $\mathbf{u}_n \neq \mathbf{u}_p$. We have shown that $n(\mathbf{e}) \neq n(\mathbf{f})$ if $\mathbf{e} \neq \mathbf{f}$, so the mapping $\mathbf{e} \to n(\mathbf{e})$ is a 1-1 function from S to a subset of the integers. It follows that S is finite or countable. $\square$

Exercises

1. Let $\mathbf{X}_1$ be the vector space of continuous complex-valued functions defined on the interval $[-1, 1]$. If $u, v \in \mathbf{X}_1$, let an inner product $(u, v)_1$ be defined by

$$(u, v)_1 = \int_{-1}^{1} u(x)v(x)^* \, dx.$$

Let $u_n(x) = x^{n-1}$, $n = 1, 2, \ldots$. Carry out the Gram-Schmidt process of Lemma 4.1 to find polynomials p_n, $n = 1, 2, 3, 4$ such that p_n is of degree $n - 1$, p_n has real coefficients, the leading coefficient is positive, and

$$(p_n, p_m)_1 = 1 \quad \text{if } n = m, \qquad (p_n, p_m)_1 = 0 \quad \text{if } n \neq m.$$

These are the first four *Legendre polynomials*.

2. Let $\mathbf{X}_2$ be the set of all continuous functions $u: \mathbb{R} \to \mathbb{C}$ such that

$$\int_{-\infty}^{\infty} |u(x)|^2 e^{-x} \, dx < \infty.$$

Show that $\mathbf{X}_2$ is a vector space. If $u, v \in \mathbf{X}_2$, show that the integral

$$(u, v)_2 = \int_{-\infty}^{\infty} u(x)v(x)e^{-x} \, dx$$

exists as an improper integral, and that this defines an inner product on $\mathbf{X}_2$. Show that there are polynomials p_n, $n = 1, 2, 3, \ldots$ such that p_n is of degree $n - 1$ and

$$(p_n, p_m)_2 = 1 \quad \text{if } n = m, \qquad (p_n, p_m)_2 = 0 \quad \text{if } n \neq m.$$

Determine the first few polynomials of such a sequence. Except for constant factors these are the *Laguerre polynomials*.

3. Show that there is a sequence $(p_n)_1^{\infty}$ of polynomials such that p_n is of degree $n - 1$ and

$$\int_{-\infty}^{\infty} |p_n(x)|^2 e^{-(1/2)x^2} \, dx = 1,$$

$$\int_{-\infty}^{\infty} p_n(x)p_m(x)e^{-(1/2)x^2} \, dx = 0 \qquad \text{if } n \neq m.$$

Except for constant factors these are the *Hermite polynomials*.

4. Suppose **H** is a finite dimensional complex Hilbert space, of dimension N. Show that there is a linear transformation U from **H** onto $\mathbb{C}^N$ such that

$$(U\mathbf{u}, U\mathbf{v}) = (\mathbf{u}, \mathbf{v}), \qquad \text{all } \mathbf{u}, \mathbf{v} \in \mathbf{H}.$$

(Hint: choose an orthonormal basis for **H**.)

5. Suppose **H** and **H**′ are two complex Hilbert spaces of dimension $N < \infty$. Show that there is a linear transformation U from **H** onto **H**′ such that

$$(U\mathbf{u}, U\mathbf{v}) = (\mathbf{u}, \mathbf{v}), \qquad \text{all } \mathbf{u}, \mathbf{v} \in \mathbf{H}.$$

6. In the space l^2 of two-sided complex sequences, let $\mathbf{e}_n$ be the sequence with entry 1 in the nth place and all other entries 0. Show that $(\mathbf{e}_n)_{-\infty}^{\infty}$ is an orthonormal basis for l^2.

7. Show that there is a linear transformation U from l^2 onto l_+^2 such that

$$(U\mathbf{x}, U\mathbf{y}) = (\mathbf{x}, \mathbf{y}), \qquad \text{all } \mathbf{x}, \mathbf{y} \in l^2$$

§5. Orthogonal expansions

Suppose **H** is a Hilbert space of dimension $N < \infty$. We know that **H** has an orthonormal basis $\{\mathbf{e}_1, \mathbf{e}_2, \ldots, \mathbf{e}_N\}$. Any element $\mathbf{u} \in \mathbf{H}$ is a linear combination

$$(1) \qquad\qquad \mathbf{u} = \sum_{n=1}^{N} a_n \mathbf{e}_n,$$

and as in §4 we see that

$$(2) \qquad\qquad a_n = (\mathbf{u}, \mathbf{e}_n).$$

It follows that if $\mathbf{u}, \mathbf{v} \in \mathbf{H}$ then

$$(3) \qquad\qquad (\mathbf{u}, \mathbf{v}) = \sum_{n=1}^{N} (\mathbf{u}, \mathbf{e}_n)(\mathbf{e}_n, \mathbf{v}).$$

In particular,

$$(4) \qquad\qquad \|\mathbf{u}\|^2 = \sum_{n=1}^{N} |(\mathbf{u}, \mathbf{e}_n)|^2.$$

The expression (1) for $\mathbf{u} \in \mathbf{H}$ with coefficients given by (2) is called the *orthogonal expansion* of **u** with respect to the orthonormal basis $\{\mathbf{e}_1, \ldots, \mathbf{e}_N\}$.

We are now in a position to carry (1)–(4) over to an infinite-dimensional separable Hilbert space.

Theorem 5.1. *Suppose **H** is a Hilbert space with an orthonormal basis* $(\mathbf{e}_n)_1^{\infty}$. *If* $\mathbf{u} \in \mathbf{H}$, *there is a unique sequence* $(a_n)_1^{\infty}$ *of scalars such that*

$$(5) \qquad\qquad \mathbf{u} = \sum_{n=1}^{\infty} a_n \mathbf{e}_n,$$

in the sense that

(6)
$$\left\| \mathbf{u} - \sum_{n=1}^{N} a_n \mathbf{e}_n \right\| \to 0 \quad as \quad N \to \infty.$$

The coefficients are given by

(7)
$$a_n = (\mathbf{u}, \mathbf{e}_n),$$

and they satisfy

(8)
$$\sum_{n=1}^{\infty} |a_n|^2 = \|\mathbf{u}\|^2.$$

More generally, if

(9)
$$\mathbf{u} = \sum_{n=1}^{\infty} a_n \mathbf{e}_n \quad and \quad \mathbf{v} = \sum_{n=1}^{\infty} b_n \mathbf{e}_n$$

then

(10)
$$(\mathbf{u}, \mathbf{v}) = \sum_{n=1}^{\infty} a_n b_n^* = \sum_{n=1}^{\infty} (\mathbf{u}, \mathbf{e}_n)(\mathbf{e}_n, \mathbf{v}).$$

Conversely, suppose $(a_n)_1^\infty$ is a sequence of scalars with the property

(11)
$$\sum_{n=1}^{\infty} |a_n|^2 < \infty.$$

Then there is a unique element $\mathbf{u} \in \mathbf{H}$ such that (5) is true.

Proof. First let us prove uniqueness. Suppose $(a_n)_1^\infty$ is a sequence of scalars such that (6) is true. Let

(12)
$$\mathbf{u}_N = \sum_{n=1}^{N} a_n \mathbf{e}_n.$$

Since the sequence $(\mathbf{e}_n)_1^\infty$ is orthonormal,

$$(\mathbf{u}_N, \mathbf{e}_n) = a_n \qquad \text{if } N \geq n.$$

Using Lemma 2.1 we get

$$a_n = \lim_{N \to \infty} (\mathbf{u}_N, \mathbf{e}_n) = (\mathbf{u}, \mathbf{e}_n).$$

Thus $(a_n)_1^\infty$ is unique.

To prove existence, set $a_n = (\mathbf{u}, \mathbf{e}_n)$. Define $\mathbf{u}_N$ by (12). Then

$$(\mathbf{u}_N, \mathbf{e}_n) = (\mathbf{u}, \mathbf{e}_n), \qquad 1 \leq n \leq N.$$

This implies that $\mathbf{u} - \mathbf{u}_N$ is orthogonal to the subspace $\mathbf{H}_N$ spanned by $\{\mathbf{e}_1, \ldots, \mathbf{e}_N\}$. Now given $\varepsilon > 0$, there is a linear combination $\mathbf{v}$ of the $\mathbf{e}_n$ such that

$$\|\mathbf{u} - \mathbf{v}\| < \varepsilon.$$

Then there is an N_0 such that $\mathbf{v} \in \mathbf{H}_N$ when $N \geq N_0$. As in the proof of Lemma 2.3, the facts that $\mathbf{u}_N \in \mathbf{H}_N$ and that $\mathbf{u} - \mathbf{u}_N \perp \mathbf{H}_N$ imply

$$\|\mathbf{u} - \mathbf{u}_N\| \leq \|\mathbf{u} - \mathbf{v}\|.$$

Thus $\mathbf{u}_N \to \mathbf{u}$. Since the e_n are orthonormal,

$$\|\mathbf{u}_N\|^2 = (\mathbf{u}_N, \mathbf{u}_N) = \sum_{n=1}^{N} |a_n|^2.$$

Thus

$$\|\mathbf{u}\|^2 = \lim \|\mathbf{u}_N\|^2 = \sum_{n=1}^{\infty} |a_n|^2.$$

More generally, suppose $\mathbf{u}$ and $\mathbf{v}$ are given by (9). Let $\mathbf{u}_N$ be defined by (12), and let $\mathbf{v}_N$ be defined in a similar way. Then by Lemma 2.1,

$$(\mathbf{u}, \mathbf{v}) = \lim (\mathbf{u}_N, \mathbf{v}_N) = \lim \sum_{n=1}^{N} a_n b_n^* = \sum_{n=1}^{\infty} a_n b_n^*.$$

Finally, suppose $(a_n)_1^\infty$ is any sequence of scalars satisfying (11). Define $\mathbf{u}_N$ by (12). All we need do is show that $(\mathbf{u}_N)_1^\infty$ is a Cauchy sequence, since we can then let $\mathbf{u}$ be its limit. But if $N > M$,

$$(13) \qquad \|\mathbf{u}_N - \mathbf{u}_M\|^2 = (\mathbf{u}_N - \mathbf{u}_M, \mathbf{u}_N - \mathbf{u}_M) = \sum_{n=M+1}^{N} |a_n|^2.$$

Since (12) is true, the right side of (13) converges to zero as $M, N \to \infty$.

It is convenient to have the corresponding statement for a Hilbert space with an orthonormal basis indexed by all integers. The proof is essentially unchanged.

Theorem 5.2. *Suppose* $\mathbf{H}$ *is a Hilbert space with an orthonormal basis* $(e_n)_{-\infty}^{\infty}$. *If* $\mathbf{u} \in \mathbf{H}$, *there is a unique two-sided sequence* $(a_n)_{-\infty}^{\infty}$ *of scalars such that*

$$(14) \qquad \mathbf{u} = \sum_{-\infty}^{\infty} a_n e_n,$$

in the sense that

$$(15) \qquad \left\| \mathbf{u} - \sum_{n=-N}^{N} a_n e_n \right\| \to 0 \quad as \quad N \to \infty.$$

The coefficients are given by

$$(16) \qquad a_n = (\mathbf{u}, e_n),$$

and they satisfy

$$(17) \qquad \sum_{n=-\infty}^{\infty} |a_n|^2 = \|\mathbf{u}\|^2.$$

More generally, if

$$\mathbf{u} = \sum_{-\infty}^{\infty} a_n \mathbf{e}_n \quad and \quad \mathbf{v} = \sum_{-\infty}^{\infty} b_n \mathbf{e}_n,$$

then

(18) $$(\mathbf{u}, \mathbf{v}) = \sum_{-\infty}^{\infty} a_n b_n^* = \sum_{-\infty}^{\infty} (\mathbf{u}, \mathbf{e}_n)(\mathbf{e}_n, \mathbf{v}).$$

Conversely, suppose $(a_n)_{-\infty}^{\infty}$ *is a sequence of scalars with the property*

$$\sum |a_n|^2 < \infty.$$

Then there is a unique $\mathbf{u} \in \mathbf{H}$ *such that* (15) *is true.*

The equations (5), (6), or (14), (15) give the *orthogonal expansion* of **u** with respect to the respective orthonormal bases. The identity (8) or (17) is called *Bessel's equality*. It implies *Bessel's inequality*:

(19) $$\sum_{n=1}^{N} |(\mathbf{u}, \mathbf{e}_n)|^2 \leq \|\mathbf{u}\|^2$$

or

(20) $$\sum_{n=-N}^{N} |(\mathbf{u}, \mathbf{e}_n)|^2 \leq \|\mathbf{u}\|^2.$$

The identity (10) or (18) is called *Parseval's identity*. The coefficients a_n given by (6) or (14) are often called the *Fourier coefficients* of **u** with respect to the respective orthonormal basis.

Exercises

1. Suppose **H** and $\mathbf{H}_1$ are two infinite-dimensional separable complex Hilbert spaces. Use Theorems 4.2, 4.3, and 5.1 to show that there is a linear transformation U from **H** onto $\mathbf{H}_1$ such that

$$(U\mathbf{u}, U\mathbf{v}) = (\mathbf{u}, \mathbf{v}), \qquad \text{all } \mathbf{u}, \mathbf{v} \in \mathbf{H}.$$

Show that U is invertible and that

$$(U^{-1}\mathbf{u}_1, U^{-1}\mathbf{v}_1) = (\mathbf{u}_1, \mathbf{v}_1), \qquad \text{all } \mathbf{u}_1, \mathbf{v}_1 \in \mathbf{H}_1.$$

Such a transformation U is called a *unitary transformation*, or a *unitary equivalence*.

2. Let $U: l_+{}^2 \to l_+{}^2$ be defined by

$$U((a_1, a_2, a_3, \ldots)) = (0, a_1, a_2, a_3, \ldots).$$

Show that U is a 1-1 linear transformation such that

$$(U\mathbf{x}, U\mathbf{y}) = (\mathbf{x}, \mathbf{y}), \qquad \text{all } \mathbf{x}, \mathbf{y} \in l_+{}^2.$$

Show that U is not onto.

§6. Fourier series

Let L^2 be the Hilbert space introduced in §1. Thus L^2 consists of each periodic distribution F which is the limit, in the sense of L^2, of a sequence $(u_n)_1^\infty \subset \mathscr{C}$, i.e.,

$$\|u_n - u_m\| \to 0,$$
$$F_{u_n} \to F \ (\mathscr{P}').$$

We can identify the space $\mathscr{C}$ of continuous periodic functions with a subspace of L^2 by identifying the *function* u with the *distribution* F_u. Then

$$\|F_u\|^2 = \|u\|^2 = \frac{1}{2\pi} \int_0^{2\pi} |u(x)|^2 \, dx.$$

In particular, we may consider the two-sided sequence of functions e_n,

$$e_n(x) = \exp(inx), \qquad n = 0, \pm 1, \pm 2, \ldots$$

as elements of L^2.

Lemma 6.1. *The sequence of functions* $(e_n)_{-\infty}^\infty$, *considered as elements of* L^2, *is an orthonormal basis for* L^2.

Proof. Clearly $\|e_n\| = 1$. If $m \neq n$, then

$$\int_0^{2\pi} e_n(x)e_m(x)^* \, dx = \int_0^{2\pi} \exp(inx - imx) \, dx$$
$$= [i(n - m)]^{-1} \exp(i(n - m)x)|_0^{2\pi} = 0.$$

Thus $(e_n)_{-\infty}^\infty$ is an orthonormal set. Now suppose $F \in L^2$. Given $\varepsilon > 0$, there is a function $u \in \mathscr{C}$ such that

$$\|F_u - F\| < \varepsilon.$$

There is a linear combination $v = \sum a_n e_n$ such that

$$|v - u| < \varepsilon.$$

Then

$$\|F_v - F\| \leq \|F_v - F_u\| + \|F_u - F\|$$
$$= \|u - v\| + \|F_u - F\|$$
$$< |v - u| + \varepsilon < 2\varepsilon. \qquad \square$$

Since $(e_n)_{-\infty}^\infty$ is an orthonormal basis, we may apply Theorem 5.2 and obtain orthogonal expansions for distributions in L^2.

Theorem 6.2. *Suppose* $F \in L^2$. *There is a unique two-sided sequence* $(a_n)_{-\infty}^\infty \subset \mathbb{C}$ *such that*

(1) $$F = \sum_{n=-\infty}^\infty a_n \exp(inx)$$

in the sense that the functions

$$(2) \qquad \sum_{n=-N}^{N} a_n e_n \to F(\mathbf{L}^2) \quad as \quad N \to \infty.$$

The coefficients are given by

$$(3) \qquad a_n = F(e_{-n}),$$

and they satisfy

$$(4) \qquad \sum_{-\infty}^{\infty} |a_n|^2 = \|F\|^2.$$

More generally, if

$$F = \sum_{-\infty}^{\infty} a_n \exp{(inx)} \quad and \quad G = \sum_{-\infty}^{\infty} b_n \exp{(inx)}$$

in the sense of (2), *then*

$$(5) \qquad (F, G) = \sum_{-\infty}^{\infty} a_n b_n^* = \sum_{-\infty}^{\infty} F(e_{-n}) G^*(e_n).$$

Conversely, suppose $(a_n)_{-\infty}^{\infty} \subset \mathbb{C}$ *and*

$$\sum_{-\infty}^{\infty} |a_n|^2 < \infty.$$

Then there is a unique $F \in \mathbf{L}^2$ *such that* (1) *is true, in the sense of* (2).

    Proof. In view of Lemma 6.1 and Theorem 5.2, we only need to verify that

$$(F, F_{e_n}) = F(e_{-n}), \qquad G(e_{-n})^* = G^*(e_n).$$

The second identity follows from the first and the definition of G^*. To prove the first, take $(u_m)_1^{\infty} \subset \mathscr{C}$,

$$u_m \to F \quad (\mathbf{L}^2).$$

Then

$$(F, F_{e_n}) = \lim (u_m, e_n) = \lim F_{u_m}(e_{-n}) = F(e_{-n}). \qquad \square$$

    The a_n in (3) are called the *Fourier coefficients* of the distribution F. The formal series on the right in (1) is called the *Fourier series* of F.

    Suppose $F = F_u$ where $u \in \mathscr{C}$. Then (2) is equivalent to

$$(6) \qquad \int_0^{2\pi} |u(x) - \sum_{n=-N}^{N} a_n \exp{(inx)}|^2 \, dx \to 0,$$

where

$$(7) \qquad a_n = \frac{1}{2\pi} \int_0^{2\pi} u(x) \exp{(-inx)} \, dx.$$

In fact, (6) and (7) remain valid when u is simply assumed to be an integrable function on $[0, 2\pi]$. In this case the a_n are called the Fourier coefficients of the *function* u, and the formal series

$$\sum_{-\infty}^{\infty} a_n \exp(inx)$$

is called the Fourier series of the function u. The fact that (6) and (7) remain valid in this case is easily established, as follows. If $u: [0, 2\pi] \to \mathbb{C}$ is integrable, then again it defines a distribution F_u by

$$F_u(v) = \frac{1}{2\pi} \int_0^{2\pi} u(x)v(x)\, dx.$$

Then an extension of the Schwarz inequality gives

$$|F_u(v)| \leq \left(\frac{1}{2\pi} \int_0^{2\pi} |u(x)|^2\, dx \right)^{1/2} \|v\| = \|u\|\, \|v\|.$$

By Exercise 2 of §2, $F_u \in \mathbf{L}^2$.

When $u \in \mathscr{C}$, it is tempting to interpret (1) for F_u as

$$u(x) = \sum_{-\infty}^{\infty} a_n \exp(inx).$$

In general, however, the series on the right may *diverge* for some values of x, and it will certainly not converge uniformly without further restrictions on u. It is sufficient to assume that u has a continuous derivative.

Lemma 6.3. *If $(a_n)_{-\infty}^{\infty} \subset \mathbb{C}$ has the property*

(8)
$$\sum_{-\infty}^{\infty} |a_n| < \infty,$$

then the functions

$$u_N = \sum_{n=-N}^{N} a_n e_n$$

converge uniformly to a function $u \in \mathscr{C}$, and $(a_n)_{-\infty}^{\infty}$ is the sequence of Fourier coefficients of u.

Proof. Since $|e_n(x)| = |\exp(inx)| = 1$ for all $x \in \mathbb{R}$, it follows from (8) that the sequence of functions $(u_N)_1^{\infty}$ is a uniform Cauchy sequence. Therefore it converges uniformly to a function $u \in \mathscr{C}$. For $N \geq m$ we have

$$(u_N, e_m) = a_m.$$

Thus

$$a_m = \lim (u_N, e_m) = (u, e_m)$$

is the mth Fourier coefficient of u. □

Theorem 6.4. *If $u \in \mathscr{C}$ and u is continuously differentiable, then the partial sums of the Fourier series of u converge uniformly to u. Thus*

$$u(x) = \sum_{-\infty}^{\infty} (u, e_n) \exp(inx), \qquad all \ x \in \mathbb{R}.$$

Moreover, if $u \in \mathscr{P}$, then the partial sums converge to u in the sense of $\mathscr{P}$.

Proof. Let $v = Du$. There is a relation between the Fourier coefficients of v and those of u. In fact integration by parts gives

(9)
$$b_n = (v, e_n) = \frac{1}{2\pi} \int_0^{2\pi} Du(x) \exp(-inx) \, dx$$

$$= \frac{in}{2\pi} \int_0^{2\pi} u(x) \exp(-inx) \, dx$$

$$= in(u, e_n) = ina_n.$$

We can apply the Schwarz inequality for sequences to show $\sum |a_n| < \infty$. In fact

$$\sum |a_n| = |a_0| + \sum_{n \neq 0} |a_n| = |a_0| + \sum_{n \neq 0} n^{-1}|b_n|$$

$$\leq |a_0| + \left(\sum_{n \neq 0} n^{-2} \right)^{1/2} \left(\sum |b_n|^2 \right)^{1/2}$$

$$= |a_0| + \left(2 \sum_{n=1}^{\infty} n^{-2} \right)^{1/2} \| Du \| < \infty.$$

By Lemma 6.3, the partial sums of the Fourier series of u converge uniformly to u.

Now suppose $u \in \mathscr{P}$, and let

$$u_N = \sum_{n=-N}^{N} a_n e_n.$$

Then (9) shows that

$$Du_N = \sum_{n=-N}^{N} ina_n e_n = \sum_{n=-N}^{N} b_n e_n$$

is the Nth partial sum of the Fourier series of Du. Similarly, $D^k u_N$ is the Nth partial sum of $D^k u$. Each $D^k u$ is continuously differentiable, so each $D^k u_N \to D^k u$ uniformly. $\square$

Fourier series expansions are very commonly written in terms of sine and cosine functions, rather than the exponential function. This is particularly natural when the function u or distribution F is real. Suppose $F \in \mathbf{L}^2$. Let

(10) $a_n = F(e_{-n}), \qquad b_n = 2F(\cos nx), \qquad c_n = 2F(\sin nx).$

Since

$$e_{-n}(x) = \cos nx - i \sin nx$$

we have

$$a_n = \tfrac{1}{2}(b_n - ic_n).$$

Also

$$b_{-n} = b_n, \qquad c_{-n} = -c_n.$$

Thus for $n > 0$,

$$a_n e_n(x) + a_{-n} e_{-n}(x) = a_n(\cos nx + i \sin nx) + a_{-n}(\cos nx - i \sin nx)$$
$$= b_n \cos nx + c_n \sin nx.$$

Then

$$\sum_{n=-N}^{N} a_n \exp(inx) = \tfrac{1}{2}b_0 + \sum_{n=1}^{N} b_n \cos nx + c_n \sin nx.$$

The formal series

(11) $$\tfrac{1}{2}b_0 + \sum_{n=1}^{\infty} b_n \cos nx + c_n \sin nx$$

is also called the *Fourier series* of F, and the coefficients b_n, c_n given by (10) are also called the *Fourier coefficients* of F. If F is real, then b_n, c_n are real, and (11) is a series of real-valued functions of x. Theorems 6.2 and 6.4 may be restated using the series (11).

Exercises

1. Find the Fourier coefficients of the following integrable functions on $[0, 2\pi]$:

(a) $u(x) = 0$, $x \in [0, \pi]$, $u(x) = 1$, $x \in (\pi, 2\pi]$.
(b) $u(x) = 0$, $x \in [0, \pi]$, $u(x) = x - \pi$, $x \in (\pi, 2\pi]$.
(c) $u(x) = |x - \pi|$.
(d) $u(x) = (x - \pi)^2$.
(e) $u(x) = x$.
(f) $u(x) = |\cos x|$.

2. Suppose $u \in \mathscr{C}$ and suppose b_n, c_n are as in (10). Show that if u is even then $c_n = 0$, all n. Show that if u is odd, then $b_n = 0$, all n. (It is convenient to integrate over $[-\pi, \pi]$ instead of $[0, 2\pi]$.) Show that if u is real then

$$\frac{1}{4} b_0{}^2 + \frac{1}{2} \sum_{n=1}^{\infty} (b_n{}^2 + c_n{}^2) = \frac{1}{2\pi} \int_0^{2\pi} u(x)^2 \, dx.$$

3. Suppose $u \in \mathscr{C}$,

$$u_N = \sum_{n=-N}^{N} (u, e_{-n})e_n.$$

Show that

$$u_N(x) = \frac{1}{2\pi} \int_0^{2\pi} D_N(x - y)u(y)\, dy,$$

where

$$D_N(x) = \sin(N + \tfrac{1}{2})x / \sin \tfrac{1}{2}x.$$

The function D_N is called the *Dirichlet kernel*. Thus

$$u_N = D_N * u.$$

4. Extend the result of Exercise 3 to the partial sums of the Fourier series of a distribution $F \in \mathbf{L}^2$.

5. For $F \in \mathscr{P}'$, define

$$a_n = F(e_{-n}), \qquad n = 0, \pm 1, \pm 2, \ldots .$$

Show that $F \in \mathbf{L}^2$ if and only if

$$\sum_{-\infty}^{\infty} |a_n|^2 < \infty.$$

6. Suppose $F \in \mathscr{P}'$ and $DF \in \mathbf{L}^2$. Show that $F = F_u$ for some continuous function u. (Hint: find the Fourier coefficients of u.)

Chapter 5

Applications of Fourier Series

§1. Fourier series of smooth periodic functions and of periodic distributions

If u is a smooth periodic function with Fourier coefficients $(a_n)_{-\infty}^{\infty}$, then we know that the sequence $(a_n)_{-\infty}^{\infty}$ uniquely determines the function u; in fact, the partial sums

$$(1) \qquad u_\wedge(x) = \sum_{-N}^{N} a_n \exp{(inx)}$$

of the Fourier series converge to u in the sense of $\mathscr{P}$. Therefore it makes sense to ask: what are necessary and sufficient conditions on a two sided sequence $(a_n)_{-\infty}^{\infty} \subset \mathbb{C}$ that it be the sequence of Fourier coefficients of a function $u \in \mathscr{P}$? The question is not hard to answer.

A sequence $(a_n)_{-\infty}^{\infty} \subset \mathbb{C}$ is said to be *of rapid decrease* if for every $r > 0$ there is a constant $c = c(r)$ such that

$$(2) \qquad |a_n| \le c|n|^{-r}, \qquad \text{all } n \neq 0.$$

Theorem 1.1. *A sequence* $(a_n)_{-\infty}^{\infty} \subset \mathbb{C}$ *is the sequence of Fourier coefficients of a function* $u \in \mathscr{P}$ *if and only if it is of rapid decrease.*

Proof. Suppose first that $u \in \mathscr{P}$ has $(a_n)_{-\infty}^{\infty}$ as its sequence of Fourier coefficients. Given $r > 0$, take an integer $k \ge r$. In proving Theorem 6.4 of Chapter 4 we noted that $(ina_n)_{-\infty}^{\infty}$ is the sequence of Fourier coefficients of Du. It follows that

$$((in)^k a_n)_{-\infty}^{\infty}$$

is the sequence of Fourier coefficients of $D^k u$. But then

$$|n|^k |a_n| \le |D^k u|,$$

which gives (2) with $c = |D^k u|$.

Conversely, suppose $(a_n)_{-\infty}^{\infty} \subset \mathbb{C}$ is a sequence which is of rapid decrease. Define functions u_N by (1). From (2) with $r = 2$ we deduce

$$\sum_{-\infty}^{\infty} |a_n| < \infty.$$

By Lemma 6.3 of Chapter 4, the functions (1) converge uniformly to $u \in \mathscr{C}$, and $(a_n)_{-\infty}^{\infty}$ are the Fourier coefficients of u. Also

$$D^k u_N = \sum_{-N}^{N} (in)^k a_n e_n,$$

and (2) with $r = k + 2$ implies

$$\sum_{-\infty}^{\infty} |n|^k |a_n| < \infty.$$

Thus each derivative $D^k u_N$ also converges uniformly as $N \to \infty$. Therefore $u \in \mathscr{P}$; see the proof of Theorem 2.1 of Chapter 3. $\square$

If F is a periodic distribution which is in $\mathbf{L}^2$, then we have defined its Fourier coefficients by

$$(3) \qquad\qquad a_n = F(e_{-n}), \qquad e_{-n}(x) = \exp(-inx).$$

Since $e_{-n} \in \mathscr{P}$, the expression in (3) makes sense for any periodic distribution F, whether or not it is in $\mathbf{L}^2$. Thus given any $F \in \mathscr{P}'$, we define its *Fourier coefficients* to be the sequence $(a_n)_{-\infty}^{\infty}$ defined by (3). We know that if all the a_n are zero, then $F = 0$ (see Chapter 3, Theorem 4.3). Therefore, $F \in \mathscr{P}'$ is uniquely determined by its Fourier coefficients, and we may ask: what are necessary and sufficient conditions on a sequence $(a_n)_{-\infty}^{\infty} \subset \mathbb{C}$ that it be the sequence of Fourier coefficients of a periodic distribution F? Again, the answer is not difficult.

A sequence $(a_n)_{-\infty}^{\infty} \subset \mathbb{C}$ is said to be *of slow growth* if there are some positive constants c and r such that

$$(4) \qquad\qquad |a_n| \le c|n|^r, \qquad \text{all } n \ne 0.$$

Theorem 1.2. *A sequence $(a_n)_{-\infty}^{\infty} \subset \mathbb{C}$ is the sequence of Fourier coefficients of a distribution $F \in \mathscr{P}'$ if and only if it is of slow growth.*

Proof. Suppose first that $F \in \mathscr{P}'$ has $(a_n)_{-\infty}^{\infty}$ as its sequence of Fourier coefficients. Recall that for some integer k, F is of order k. Thus for $u \in \mathscr{P}$,

$$|F(u)| \le c(|u| + |Du| + \cdots + |D^k u|).$$

With $u = e_{-n}$ this means

$$|a_n| = |F(u)| \le c(|1 + |n| + \cdots + |n|^k) < 2c|n|^{k+1}.$$

Thus $(a_n)_{-\infty}^{\infty}$ is of slow growth.

Conversely, suppose $(a_n)_{-\infty}^{\infty} \subset \mathbb{C}$ is a sequence which is of slow growth. Then there is an integer $k > 0$ such that

$$(5) \qquad\qquad |a_n| \le c|n|^{k-2}, \qquad n \ne 0.$$

Let $b_0 = 0$ and

$$b_n = (in)^{-k} a_n, \qquad n \ne 0.$$

Let

$$v_N = \sum_{-N}^{N} b_n e_n.$$

From (5) we get

$$\sum_{-\infty}^{\infty} |b_n| < \infty.$$

Therefore v_N converges uniformly to $v \in \mathscr{C}$. Let

$$F = D^k F_v + F_f, \qquad f = a_0 e_0.$$

This is a distribution; we claim that its Fourier coefficients are the $(a_n)^\infty_{-\infty}$. In fact, for $n \neq 0$,

$$\begin{aligned}
F(e_{-n}) &= D^k F_v(e_{-n}) = F_v((-D)^k e_{-n}) \\
&= (in)^k F_v(e_{-n}) = (in)^k b_n = a_n.
\end{aligned}$$

Also

$$F(e_0) = D^k F_v(e_0) + F_f(e_0) = 0 + a_0. \qquad \square$$

In the course of the preceding proof we gave a second proof of the characterization theorem for periodic distributions, Theorem 6.1 of Chapter 3. In fact, the whole theory of $\mathscr{P}$ and $\mathscr{P}'$ in Chapter 3 can be derived from the point of view of Fourier series. We shall do much of such a derivation in this section and the next. An important feature of such a program is to express the action of $F \in \mathscr{P}'$ on $u \in \mathscr{P}$ in terms of the respective sequences of Fourier coefficients.

Theorem 1.3. *Suppose $F \in \mathscr{P}'$ has $(a_n)^\infty_{-\infty}$ as its sequence of coefficients, and suppose $u \in \mathscr{P}$ has $(b_n)^\infty_{-\infty}$ as its sequence of Fourier coefficients. Then*

$$(6) \qquad F(u) = \sum_{-\infty}^{\infty} a_n b_{-n} = \sum_{-\infty}^{\infty} a_{-n} b_n.$$

Proof. Let

$$u_N(x) = \sum_{-N}^{N} b_n \exp(inx).$$

We know

$$u_N \to u \ (\mathscr{P}).$$

Therefore

$$F(u) = \lim F(u_N).$$

But

$$F(u_N) = \sum_{-N}^{N} b_n F(e_n) = \sum_{-N}^{N} a_{-n} b_n = \sum_{-N}^{N} a_n b_{-n}. \qquad \square$$

Implicit in the proof of Theorem 1.3 is the proof that the series in (6) converges. A more direct proof uses the criteria in Theorem 1.1 and 1.2. In fact,

$$\begin{aligned}
|a_n| &\leq c|n|^r, & n \neq 0, \\
|b_n| &\leq c'|n|^{-r-2}, & n \neq 0
\end{aligned}$$

and convergence follows.

Corollary 1.4. *If $F \in \mathscr{P}'$ has Fourier coefficients $(a_n)_{-\infty}^{\infty}$, and if F_N is the distribution defined by the function*

$$\sum_{-N}^{N} a_n \exp (inx),$$

then

$$F_N \to F (\mathscr{P}').$$

Proof. With u, u_N as in Theorem 1.3,

$$F_N(u) = F(u_N) \to F(u). \qquad \qquad \square$$

Exercises

1. Compute the Fourier coefficients of δ and $D^k\delta$.

2. If $F \in \mathscr{P}'$ has Fourier coefficients $(a_n)_{-\infty}^{\infty}$, compute the Fourier coefficients of

$$T_t F, \qquad F^*, \qquad F^\sim.$$

3. Give necessary and sufficient conditions on the Fourier coefficients $(a_n)_{-\infty}^{\infty}$ of a distribution F that F be real; or even; or odd.

4. Suppose $(\varphi_m)_1^{\infty} \subset \mathscr{C}$ is an approximate identity. Let

$$(a_{m,n})_{n=-\infty}^{\infty}$$

be the sequence of Fourier coefficients of φ_m. Show that

$$|a_{m,n}| \leq 1, \qquad m \geq 1, \qquad n = 0, \pm 1, \pm 2, \ldots;$$
$$\lim_{m \to \infty} a_{m,n} = 1, \qquad \text{all } n.$$

§2. Fourier series, convolutions, and approximation

Recall that if F, $G \in \mathscr{P}'$ and $u \in \mathscr{P}$, the convolutions $F * u$ and $F * G$ are defined by

$$F * u(x) = F(T_x \tilde{u}),$$
$$(F * G)(u) = F(G^\sim * u).$$

We want to compute the Fourier coefficients of the convolutions in terms of the Fourier coefficients of F, G, and u.

Theorem 2.1. *Suppose $F \in \mathscr{P}'$ has Fourier coefficients $(a_n)_{-\infty}^{\infty}$; suppose $G \in \mathscr{P}'$ has Fourier coefficients $(b_n)_{-\infty}^{\infty}$; and suppose $u \in \mathscr{P}$ has Fourier coefficients $(c_n)_{-\infty}^{\infty}$. Then $F * u$ has Fourier coefficients $(a_n c_n)_{-\infty}^{\infty}$ and $F * G$ has Fourier coefficients $(a_n b_n)_{-\infty}^{\infty}$.*

Proof. Note that

$$(T_x \tilde{e}_n)(y) = e_{-n}(y - x) = e_n(x)e_{-n}(y).$$

Therefore

$$(F * e_n)(x) = e_n(x)F(e_{-n}) = a_n e_n(x),$$

and

$$F * \left(\sum_{-N}^{N} c_n e_n \right) = \sum_{-N}^{N} a_n c_n e_n.$$

Taking the limit as $N \to \infty$, we find that $F * u$ has Fourier coefficients $(a_n c_n)_{-\infty}^{\infty}$.

Now

$$(G^{\sim} * e_n)(x) = G^{\sim}(T_x \tilde{e}_n) = e_n(x)G^{\sim}(\tilde{e}_n)$$
$$= e_n(x)G(e_n) = b_{-n}e_n(x).$$

Therefore

$$(F * G)(e_{-n}) = F(G^{\sim} * e_{-n}) = b_n F(e_{-n}) = a_n b_n. \qquad \square$$

Using Theorem 2.1 and Theorems 1.1 and 1.2, we may easily give a second proof that $F * u \in \mathscr{P}$. Similarly, if $u \in \mathscr{P}$ and $G = F_u$, then

$$F * G = F_v, \quad \text{where} \quad v = F * u;$$

in fact $F * G$ and $F * u$ have the same Fourier coefficients.

The approximation theorems of Chapter 3 may be proved using Theorem 2.1 and the following two general approximation theorems.

Theorem 2.2. *Suppose* $(u_m)_1^{\infty} \subset \mathscr{P}$. *Suppose the Fourier coefficients of* u_m *are*

$$(a_{m,n})_{n=-\infty}^{\infty}.$$

Suppose that for each $r > 0$ *there is a constant* $c = c(r)$ *such that*

$$|a_{m,n}| \le c|n|^{-r}, \quad \text{all } m, \text{ all } n \ne 0.$$

Suppose, finally, that for each n,

$$a_{m,n} \to a_n \quad \text{as} \quad m \to \infty.$$

Then $(a_n)_{-\infty}^{\infty}$ *is the sequence of Fourier coefficients of a function* $u \in \mathscr{P}$. *Moreover,*

$$u_m \to u \ (\mathscr{P}).$$

Proof. The conditions imply that also

$$|a_n| \le c|n|^{-r}, \quad n \ne 0.$$

Thus $(a_n)_{-\infty}^{\infty}$ is the sequence of Fourier coefficients of a function $u \in \mathscr{P}$. Given $\varepsilon > 0$, choose N so large that

$$c(2) \sum_{n=N}^{\infty} n^{-2} < \varepsilon.$$

Choose M so large that $m \geq M$ implies

$$\sum_{-N}^{N} |a_{m,n} - a_n| < \varepsilon.$$

It follows that if $m \geq M$ then

$$\sum_{-\infty}^{\infty} |a_{m,n} - a_n| < 5\varepsilon.$$

Since

$$u = \sum a_n e_n, \qquad u_m = \sum a_{m,n} e_n,$$

this implies

$$|u_m - u| < 5\varepsilon \qquad \text{if } m \geq M.$$

Thus $u_m \to u$ uniformly. A similar argument shows each $D^k u_m \to D^k u$. □

Theorem 2.3. *Suppose* $(F_m)_1^\infty \subset \mathscr{P}'$. *Suppose the Fourier coefficients of* F_m *are*

$$(a_{m,n})_{n=-\infty}^\infty.$$

Suppose that for some $r > 0$ there is a constant c such that

$$|a_{m,n}| \leq c|n|^r, \qquad \text{all } m, \text{ all } n \neq 0.$$

Suppose, finally, that for each n,

$$a_{m,n} \to a_n \quad \text{and} \quad m \to \infty.$$

Then $(a_n)_{-\infty}^\infty$ is the sequence of Fourier coefficients of a distribution $F \in \mathscr{P}'$. Moreover,

$$F_m \to F \ (\mathscr{P}').$$

Proof. The conditions imply that also

$$|a_n| \leq c|n|^r, \qquad \text{all } n \neq 0.$$

Thus $(a_n)_{-\infty}^\infty$ is the sequence of Fourier coefficients of a distribution $F \in \mathscr{P}'$. Take an integer $k \geq r + 2$, and let

$$\begin{aligned} b_{m,0} &= b_0 = 0, \\ b_{m,n} &= (in)^{-k} a_{m,n}, \qquad n \neq 0 \\ b_n &= (in)^{-k} a_n, \qquad n \neq 0. \end{aligned}$$

As in the proof of Theorem 1.2, $(b_{m,n})_{n=-\infty}^\infty$ is the sequence of Fourier coefficients of a function $v_m \in \mathscr{C}$, with

$$F_m = D^k F_{v_m} + F_{f_m}, \qquad f_m = a_{m,0} e_0.$$

Similarly, $(b_n)_{-\infty}^\infty$ is the sequence of Fourier coefficients of $v \in \mathscr{C}$ with

$$F = D^k F_v + F_f, \qquad f = a_0 e_0.$$

The hypotheses of the theorem imply

$$|b_{m,n}| \leq c|n|^{-2}, \qquad n \neq 0,$$
$$b_{m,n} \to b_n \quad \text{as} \quad m \to \infty.$$

As in the proof of Theorem 2.1, these conditions imply that

$$v_m \to v \quad \text{uniformly as} \quad m \to \infty.$$

Also,

$$a_{m,0} \to a_0.$$

It follows that $F_m \to F$ $(\mathscr{P}')$. □

Exercises

1. Suppose $(\varphi_m)_1^\infty \subset \mathscr{C}$ is an approximate identity. Use the theorems of this section and Exercise 4 of §1 to prove that:

$$\varphi_m * u \to u \ (\mathscr{P}) \qquad \text{if } u \in \mathscr{P};$$
$$F * \varphi_m \to F \ (\mathscr{P}') \qquad \text{if } F \in \mathscr{P}'.$$

2. State and prove a theorem for $\mathbf{L}^2$ which is analogous to Theorem 2.2 for $\mathscr{P}$ and Theorem 2.3 for $\mathscr{P}'$.

3. Use the result of Exercise 2 to show that if $(\varphi_m)_1^\infty \subset \mathscr{C}$ is an approximate identity and $F \in \mathbf{L}^2$, then

$$F * \varphi_m \to F \ (\mathbf{L}^2).$$

4. Prove the converse of Theorem 2.2: if $u_m \to u$ $(\mathscr{P})$ then the Fourier coefficients satisfy the hypotheses of Theorem 2.2.

§3. The heat equation: distribution solutions

Many physical processes are approximately described by a function u, depending on time and on position in space, which satisfies a type of partial differential equation called a *heat equation* or *diffusion equation*. The simplest case is the following. Find $u(x, t)$, a continuous function defined for $x \in [0, \pi]$ and for $t \geq 0$, satisfying the equation

(1) $$\frac{\partial}{\partial t} u(x, t) = \kappa \left(\frac{\partial}{\partial x}\right)^2 u(x, t), \qquad x \in (0, \pi), \qquad t > 0,$$

the *initial condition*

(2) $$u(x, 0) = g(x), \qquad x \in [0, \pi],$$

and one of the following two sets of *boundary conditions*:

(3) $u(0, t) = u(\pi, t) = 0, \qquad t \geq 0;$

(3)' $\dfrac{\partial}{\partial x} u(0, t) = \dfrac{\partial}{\partial x} u(\pi, t) = 0, \qquad t > 0.$

The function u describes the temperature distribution in a thin homogeneous metal rod of length π. The number x represents the distance of the point P on the rod from one end of the rod, the number t represents the time, and the number $u(x, t)$ the temperature at the point P at time t. Equation (1) expresses the assumption that the rod is in an insulating medium, with no heat gained or lost except possibly at the ends. The constant $\kappa > 0$ is proportional to the thermal conductivity of the metal, and we may assume units are chosen so that $\kappa = 1$. Equation (2) expresses the assumption that the temperature distribution is known at time $t = 0$. Equation (3) expresses the assumption that the ends of the rod are kept at the constant temperature 0, while the alternative equation (3)' applies if the ends are assumed insulated. Later we shall sketch the derivation of Equation (1) and indicate some other physical processes it describes.

Let us convert the two problems (1), (2), (3) and (1), (2), (3)' into a single problem for a function periodic in x. Note that if (2) and (3) are both to hold, we should have

(4) $g(0) = g(\pi) = 0.$

Let $g(-x) = -g(x)$, $x \in (-\pi, 0)$. Then g has a unique extension to all of $\mathbb{R}$ which is odd and periodic (period 2π). Because of (4) the resulting function is still continuous. Suppose u were a function defined for all $x \in \mathbb{R}$ and $t \geq 0$, periodic in x, and satisfying (1) for all $x \in \mathbb{R}$, $t > 0$. Then $u(x, 0) = g(x)$ is odd. If g is smooth, then also

$$\left(\frac{\partial}{\partial x}\right)^2 u(x, 0) = D^2 g(x)$$

is odd, and we might expect that u is odd as a function of x for each $t \geq 0$. If this is so, then necessarily (3) is true.

Similarly, if (2) and (3)' are both to hold, we would expect that if g is smooth then

$$Dg(0) = Dg(\pi) = 0.$$

In this case, let $g(-x) = g(x)$, $x \in (-\pi, 0)$. Then g has a unique extension to all of $\mathbb{R}$ which is even and periodic. If g is of class C^1 on $[0, \pi]$, the extension is of class C^1 on $\mathbb{R}$. Again, if u were a solution of (1), (2) which is periodic in x for all t, we might expect u to be even for all t. Then (3)' is necessarily true.

The above considerations suggest that we replace the two problems above by the single problem: find u defined for $x \in \mathbb{R}$, $t \geq 0$, such that u is periodic in x,

(5) $$\frac{\partial}{\partial t} u(x, t) = \left(\frac{\partial}{\partial x}\right)^2 u(x, t), \qquad x \in \mathbb{R}, \qquad t > 0,$$

(6) $$u(x, 0) = g(x), \qquad x \in \mathbb{R},$$

where $g \in \mathscr{C}$ is given.

It is convenient and useful to ask for solutions of an analogous problem for periodic *distributions*. We formulate this more general problem as follows. Suppose that to each t in some interval $(a, b) \in \mathbb{R}$ we have assigned a distribution $F_t \in \mathscr{P}'$. If $s \in (a, b)$, $G \in \mathscr{P}'$, and

$$(t - s)^{-1}(F_t - F_s) \to G \; (\mathscr{P}')$$

as $t \to s$, it is natural to consider G as the derivative of F_t with respect to t at $t = s$. We do so, and write

$$G = \frac{d}{dt} F_t \bigg|_{t=s}.$$

Our formulation of the problem for distribution is as follows: given $G \in \mathscr{P}'$, find distributions $F_t \in \mathscr{P}'$ for each $t > 0$, such that

(7) $$\frac{d}{dt} F_t \bigg|_{t=s} = D^2 F_s, \qquad \text{all } s > 0,$$

(8) $$F_t \to G \; (\mathscr{P}') \quad \text{as} \quad t \to 0.$$

Theorem 3.1. *For each $G \in \mathscr{P}'$ there is a unique family $(F_t)_{t>0} \subset \mathscr{P}'$ such that (7) and (8) hold. For each $t > 0$, F_t is a function $u_t(x) = u(x, t)$ which is infinitely differentiable in both variables and satisfies (5).*

Proof. Let us prove uniqueness first. Suppose $(F_t)_{t>0}$ is a solution of (7), (8). Each F_t has Fourier coefficients $(a_n(t))_{-\infty}^{\infty}$,

$$a_n(t) = F_t(e_{-n}), \qquad t > 0, n = 0, \pm 1, \ldots.$$

Then

$$(t - s)^{-1}[a_n(t) - a_n(s)] \to D^2 F_s(e_{-n}) = F_s(D^2 e_{-n}) = -n^2 F_s(e_{-n})$$

as $t \to s$. In other words,

(9) $$Da_n(t) = -n^2 a_n(t), \qquad t > 0.$$

As $t \to 0$ we have

(10) $$a_n(t) = F_t(e_{-n}) \to G(e_{-n}) = b_n.$$

The *unique* function $a(t)$, $t > 0$ which satisfies (9) and (10) is

(11) $$a_n(t) = b_n \exp(-n^2 t).$$

This shows that $a_n(t)$ are uniquely determined, so the distributions F_t are uniquely determined.

To show *existence*, we want to show that the $a_n(t)$ defined by (11) are the Fourier coefficients of a smooth function for $t > 0$. Recall that if $y > 0$ then

$$e^y = \sum (n!)^{-1} y^n > (m!)^{-1} y^m$$

so

$$e^{-y} < m!\, y^{-m}.$$

Therefore

$$|a_n(t)| \le |b_n|\, m!\, n^{-2m} \cdot t^{-m}, \qquad m = 0, 1, 2, \ldots.$$

But

$$|b_n| \le c|n|^k, \qquad \text{all } n \ne 0$$

for some c and k. Therefore

$$|a_n(t)| \le cm!\, |n|^{k-2m} \cdot t^{-m}, \qquad m = 0, 1, 2, \ldots.$$

It follows that for each $t > 0$, $(a_n(t))^\infty_{-\infty}$ is the sequence of Fourier coefficients of a function $u_t \in \mathcal{P}$. Then u_t is the uniform limit of the functions

$$u_N(x, t) = \sum_{-N}^{N} a_n(t) \exp{(inx)}$$

$$= \sum_{-N}^{N} b_n \exp{(inx - n^2 t)}.$$

Then

$$\left(\frac{\partial}{\partial t}\right)^l \left(\frac{\partial}{\partial x}\right)^r u_N(x, t) = \sum_{-N}^{N} b_n(-n^2)^l (in)^r \exp{(inx - n^2 t)}$$

$$= \sum_{-N}^{N} a_{n,l,r}(t) \exp{(inx)}.$$

As above,

$$|a_{n,l,r}(t)| \le cm!\, |n|^{k+2l+r-2m} \cdot t^{-m}, \qquad m = 0, 1, 2, \ldots.$$

It follows that each partial derivative of u_N converges as $N \to \infty$, uniformly for $x \in \mathbb{R}$ and $t \ge \delta > 0$. Thus

$$u(x, t) = u_t(x)$$

is smooth for $x \in \mathbb{R}$, $t > 0$. For each N, u_N satisfies (5). Therefore u satisfies (5).

Let F_t be the distribution determined by u_t, $t > 0$. Thus the Fourier coefficients of F_t are the same as those of u_t.

It follows from (11) that

$$|a_n(t)| \le |b_n|, \qquad a_n(t) \to b_n \quad \text{as} \quad t \to 0.$$

By Theorem 2.3, therefore, (8) is satisfied. Finally, each partial derivative of u is bounded on the region $x \in \mathbb{R}$, $t \geq \delta > 0$. It follows from this and the mean value theorem that

$$(t - s)^{-1}(u_t - u_s) \to \frac{\partial}{\partial t} u_s$$

uniformly as $t \to s > 0$. Therefore (7) also holds.   □

Exercises

1. In Theorem 3.1, suppose G is even or odd. Show that each F_t is respectively even or odd. Show that if G is real, then each F_t is real.

2. Discuss the behavior of F_t as $t \to \infty$.

3. Formulate the correct conditions on the function u if it represents the temperature in a rod with one end insulated and the other held at constant temperature.

4. In Theorem 3.1, suppose $G = F_w$, the distribution defined by a function $w \in \mathscr{P}$. Show that the functions $u_t \to w$ ($\mathscr{P}$) as $t \to 0$.

5. Let

$$g_t(x) = \sum_{n = -\infty}^{\infty} \exp(-n^2 t + inx), \qquad t > 0, \qquad x \in \mathbb{R}.$$

Show that $g_t \in \mathscr{P}$. In Theorem 3.1, show that

$$u_t = G * g_t.$$

6. With g_t as in Exercise 5, show that

$$g_t * g_s = g_{t+s}.$$

7. The *backwards heat equation* is the equation (1) considered for $t \leq 0$, with initial (or "final") condition (2). Is it reasonable to expect that solutions for this problem will exist? Specifically, given $G \in \mathscr{P}'$, will there always be a family of distributions $(F_t)_{t<0} \subset \mathscr{P}'$ such that

$$\frac{d}{dt} F_t \Big|_{t=s} = D^2 F_s, \qquad \text{all } s < 0,$$

$$F_t \to G \ (\mathscr{P}') \quad \text{as} \quad t \to 0?$$

8. The *Schrödinger equation* (simplest form) is the equation

$$\frac{\partial u}{\partial t} = i \frac{\partial^2 u}{\partial x^2}.$$

Consider the corresponding problem for periodic distributions: given $G \in \mathscr{P}'$, find a family $(F_t)_{t>0}$ of periodic distributions such that

$$\frac{d}{dt} F_t \Big|_{t=s} = i D^2 F_s, \qquad \text{all } s > 0,$$

$$F_t \to G \ (\mathscr{P}') \quad \text{as} \quad t \to 0.$$

Discuss the existence and uniqueness of solutions to this problem.

§4. The heat equation: classical solutions; derivation

Let us return to the problem given at the beginning of the last section: find $u(x, t)$, continuous for $x \in [0, \pi]$ and $t \geq 0$, and satisfying

(1) $$\frac{\partial}{\partial t} u(x, t) = \left(\frac{\partial}{\partial x}\right)^2 u(x, t), \qquad x \in (0, \pi), \qquad t > 0;$$

(2) $$u(x, 0) = g(x), \quad \text{where} \quad g \in C[0, \pi] \text{ is given};$$

(3) $$u(0, t) = u(\pi, t) = 0, \quad t \geq 0$$

or

(3)′ $$\frac{\partial}{\partial x} u(0, t) = \frac{\partial}{\partial x} u(\pi, t) = 0, \qquad t > 0.$$

Such a function u is called a *classical solution* of the problem (1), (2), (3) or (1), (2), (3)′, in contrast to the distribution solution for the periodic problem given by Theorem 3.1. In this section we complete the discussion by showing that a classical solution exists and is unique, and that it is given by the distribution solution. We consider the problem (1), (2), (3), and leave the problem (1), (2), (3)′ as an exercise.

Given $g \in C[0, \pi]$ with $g(0) = g(\pi) = 0$ (so that (3) is reasonable), extend g to be an odd periodic function in $\mathscr{C}$, and let $G = F_g$. By Theorem 3.1, there is a function $u(x, t) = u_t(x)$ which is smooth in x, t for $x \in \mathbb{R}$, $t > 0$, satisfies (1) for all such x, t, and which converges to G in the sense of $\mathscr{P}'$ as $t \to 0$. By Exercise 1 of §3, u is odd as a function of x for each $t > 0$. Since u is also periodic as a function of x, this implies that (3) holds when $t > 0$. *If we knew that $u_t \to g$ uniformly as $t \to 0$*, it would follow that the restriction of u to $0 \leq x \leq \pi$ is a classical solution of (1), (2), (3). Note that this is true when (the extension of) g is smooth: see Exercise 4 of §3. Everything else we need to know follows from the *maximum principle* stated in the following theorem.

Theorem 4.1. *Suppose u is a real-valued classical solution of* (1), (2). *Then for each $T > 0$, the maximum value of $u(x, t)$ in the rectangle*

$$0 \leq x \leq \pi, \qquad 0 \leq t \leq T$$

is attained on one of the three edges $t = 0$, $x = 0$, or $x = \pi$.

Proof. Given $\varepsilon > 0$, let

$$v(x, t) = u(x, t) - \varepsilon t.$$

It is easy to see that the maximum value of v is attained on one of the edges in question. Otherwise, it would be attained at (x_0, t_0), where

$$x_0 \in (0, \pi), \qquad t_0 > 0.$$

For v to be maximal here, we must have

$$\frac{\partial}{\partial t} v(x_0, t_0) \geq 0.$$

But then

$$\left(\frac{\partial}{\partial x}\right)^2 v(x_0, t_0) = \left(\frac{\partial}{\partial x}\right)^2 u(x_0, t_0) = \frac{\partial}{\partial t} u(x_0, t_0) = \frac{\partial}{\partial t} v(x_0, t_0) + \varepsilon$$

$$\geq \varepsilon > 0,$$

so x_0 cannot be a maximum for $v(x, t_0)$ on $[0, 2\pi]$.

Now u and v differ by at most εT on the rectangle. Therefore for any (x, t) in the rectangle,

$$u(x, t) \leq M + 2\varepsilon T,$$

where M is the maximum of u for $t = 0$, $x = 0$, or $x = \pi$. Since this is true for every $\varepsilon > 0$, the conclusion follows. $\square$

Theorem 4.2. *For each continuous g with $g(0) = g(\pi) = 0$, there is a unique classical solution of problem* (1), (2), (3).

Proof. Note that u is a classical solution with initial values given by g if and only if the real and imaginary parts of u are classical solutions with initial values given by the real and imaginary parts of g, respectively. Therefore we may assume that g and u are real-valued. Applying Theorem 4.1 we see that

$$u(x, t) \leq |g|, \qquad \text{all } x \in [0, \pi], \qquad t \geq 0.$$

Applying Theorem 4.1 to $-u$, which is a solution with initial values given by $-g$, we get

$$-u(x, t) \leq |g|, \qquad \text{all } x, t.$$

Thus

$$|u(x, t)| \leq |g|, \qquad \text{all } x, t.$$

This proves uniqueness.

To prove existence, let g be extended so as to be odd and periodic. Let $(\varphi_m)_1^\infty \subset \mathscr{P}$ be an approximate identity. Let

$$g_m = \varphi_m * g \in \mathscr{P}.$$

We can choose φ_m to be even, so that g_m is odd. Let u_m be the distribution solution given by Theorem 3.1 for g_m as initial value, and let u be the distribution solution with g as initial value. Then we know $u_m(x, t) \to g_m(x)$ uniformly with respect to x as $t \to 0$, so we may consider u_m as continuous for $t \geq 0$, $x \in \mathbb{R}$. Moreover, since $g_m \to g$ uniformly as $m \to \infty$, it follows that $u_m(x, t) \to u(x, t)$, at least in the sense of $\mathscr{P}'$, for each $t > 0$. On the other hand, since $|g_m - g| \to 0$ we find that for $x \in [0, \pi]$

$$|u_m(x, t) - u_p(x, t)| \to 0$$

as $m, p \to \infty$, uniformly in x and t, $t > 0$. Thus we must have $u_m \to u$ uniformly. It follows that u has a continuous extension to $t = 0$, and therefore that u (restricted to $x \in [0, \pi]$) is a classical solution. $\square$

A second proof of the uniform convergence of u to g as $t \to 0$ is sketched in the exercises below.

The heat equation, (1), may be derived as follows. Again we consider a homogeneous thin metal rod in an insulating medium. Imagine the rod divided into sections of length ε, and suppose x is the coordinate of the midpoint of one section. We consider only this and the two adjacent sections, and approximate the temperature distribution at time t by assuming u to be constant in each section. The rate of flow of *heat* from the section centered at $x + \varepsilon$ to that centered at x is proportional to the temperature difference $u(x + \varepsilon, t) - u(x, t)$, and inversely proportional to the distance ε. The *temperature* in each section is the amount of heat divided by the volume, and the volume is proportional to ε. Considering also the heat flow from the section centered at $x - \varepsilon$ to that centered at x, we get an approximate expression for the rate of change of temperature at (x, t):

$$\varepsilon \frac{\partial}{\partial t} u(x, t) \approx \kappa \varepsilon^{-1}[u(x + \varepsilon, t) - u(x) + u(x - \varepsilon, t) - u(x)]$$

or

$$\frac{\partial}{\partial t} u(x, t) \approx \kappa \varepsilon^{-2}[u(x + \varepsilon, t) - u(x - \varepsilon, t) - 2u(x)].$$

Consider the expression on the right as a function of ε and let $\varepsilon \to 0$. Two applications of L'Hôpital's rule give

(4) $$\frac{\partial}{\partial t} u(x, t) = \kappa \left(\frac{\partial}{\partial x}\right)^2 u(x, t).$$

Note that essentially the same reasoning applies to the following general situation. A (relatively) narrow cylinder contains a large number of individual objects which move rather randomly about. The random motion of each object is assumed symmetric in direction (left or right is equally likely) and essentially independent of position in the cylinder, past motion, or the presence of the other objects. As examples one can picture diffusion of molecules or dye in a tube of water kicked about by thermal motion of the water molecules, or the late stages of a large cocktail party in a very long narrow room. If $u(x, t)$ represents the density of the objects near the point x at time t, then equation (4) arises again. Boundary conditions like (3)' correspond to the ends of the cylinder being closed, while those like (3) correspond to having one way doors at the ends, to allow egress but not ingress.

Exercises

1. Suppose $G = F_w$, $w \in \mathscr{P}$, and suppose $(u_t)_{t>0}$ is the family of functions in Theorem 3.1. Show that if $w \geq 0$, then each u_t is ≥ 0.

2. As in Exercise 5 of §3, let

$$g_t(x) = \sum_{-\infty}^{\infty} \exp(-n^2 t + inx), \qquad t > 0.$$

Show that

$$\frac{1}{2\pi} \int_0^{2\pi} g_t(x)\, dx = 1.$$

Show that if $w \in \mathscr{P}$ and $w \geq 0$, then $g_t * w \geq 0$. Show, by using any approximate identity in $\mathscr{P}$, that $g_t \geq 0$.

3. Show that $(g_t)_{t>0}$ is an approximate identity as $t \to 0$, i.e., (in addition to the conclusions of Exercise 2) for each $0 < \delta < \pi$,

$$\lim_{t \to 0} \int_\delta^{2\pi - \delta} g_t(x)\, dx = 0.$$

(Hint: choose $w \in \mathscr{P}$ such that $w \geq 0$, $w(0) = 0$, and $w(x) = 1$ for $\delta \leq x \leq 2\pi - \delta$. Then consider $g_t * w(0)$.)

4. Use Exercise 3 to show that if $w \in \mathscr{C}$, $G = F_w$ and $(u_t)_{t>0}$ is the family of functions given in Theorem 3.1, then $u_t \to w$ uniformly as $t \to 0$.

5. In a situation in which heat is being supplied to or drawn from a rod with ends at a fixed temperature, one is led to the problem

$$\frac{\partial}{\partial t} u(x, t) = \left(\frac{\partial}{\partial x}\right)^2 u(x, t) + f(x, t), \qquad x \in (0, \pi), \qquad t > 0,$$

$$u(x, 0) = g(x), \qquad u(0, t) = u(\pi, t) = 0.$$

Formulate and solve the corresponding problem for periodic distributions, and discuss existence and uniqueness of classical solutions.

§5. The wave equation

Another type of equation which is satisfied by the functions describing many physical processes is the wave equation. The simplest example occurs in connection with a vibrating string. Consider a taut string of length π with endpoints fixed at the same height, and let $u(x, t)$ denote the vertical displacement of the string at the point with coordinate x, at time t. If there are no external forces, the function u is (approximately) a solution of the equation

$$(1) \qquad \left(\frac{\partial}{\partial t}\right)^2 u(x, t) = c^2 \left(\frac{\partial}{\partial x}\right)^2 u(x, t), \qquad x \in (0, \pi), \qquad t > 0.$$

Here c is a constant depending on the tension and properties of the string. The condition that the endpoints be fixed is

$$(2) \qquad u(0, t) = u(\pi, t) = 0, \qquad t \geq 0.$$

To complete the determination of u it is enough to know the position and velocity of each point of the string at time $t = 0$:

$$(3) \qquad u(x, 0) = g(x), \qquad \frac{\partial}{\partial t} u(x, 0) = h(x), \qquad x \in [0, \pi].$$

We shall discuss the derivation of (1) later in this section.

As in the case of the heat equation we begin by formulating a corresponding problem for periodic distributions and solving it. The conditions (2) suggest that we extend g, h to be odd and periodic and look for a solution periodic in x. The procedure is essentially the same as in §3.

Theorem 5.1. *For each G, $H \in \mathcal{P}'$ there is a unique family $(F_t)_{t>0} \subset \mathcal{P}'$ with the following properties:*

$$\frac{d}{dt} F_t \bigg|_s \qquad \text{exists for each } s > 0,$$

$$(4) \qquad \left(\frac{d}{dt}\right)^2 F_t \bigg|_s = D^2 F_s \qquad \text{all } s > 0,$$

$$(5) \qquad F_t \to G \ (\mathcal{P}') \quad \text{as} \quad t \to 0,$$

$$(6) \qquad \frac{d}{dt} F_t \bigg|_s \to H \ (\mathcal{P}') \quad \text{as} \quad s \to 0.$$

Proof. Let $(b_n)_{-\infty}^{\infty}$ be the sequence of Fourier coefficients of G, and let $(c_n)_{-\infty}^{\infty}$ be the sequence of Fourier coefficients of H. If $(F_t)_{t>0}$ is such a family of distributions, let the Fourier coefficients of F_t be

$$(a_n(t))_{-\infty}^{\infty}.$$

As in the proof of Theorem 3.1, conditions (4), (5), (6) imply that a_n is twice continuously differentiable for $t > 0$, a_n and Da_n have limits at $t = 0$, and

$$D^2 a_n(t) = -n^2 a_n(t),$$
$$a_n(0) = b_n, \qquad Da_n(0) = c_n.$$

The unique function satisfying these conditions (see §6 of Chapter 2) is

$$(7) \qquad a_n(t) = b_n \cos nt + n^{-1} c_n \sin nt, \qquad n \neq 0,$$

$$(8) \qquad a_0(t) = b_0 + c_0 t.$$

Thus we have proved uniqueness. On the other hand, the functions (7), (8) satisfy

$$(9) \qquad |a_n(t)| \leq |b_n| + |n|^{-1} |c_n|, \qquad n \neq 0,$$

$$(10) \qquad a_n(t) \to b_n \quad \text{as} \quad t \to 0, \qquad \text{all } n;$$

$$(11) \qquad |Da_n(t)| \leq |nb_n| + |c_n|, \qquad \text{all } n,$$

$$(12) \qquad Da_n(t) \to c_n \quad \text{as} \quad t \to 0, \qquad \text{all } n.$$

It follows from (9) and Theorem 1.2 that $(a_n(t))_{-\infty}^{\infty}$ is the sequence of Fourier coefficients of a distribution F_t. It follows from (10) and Theorem 2.3 that (5) is true. It follows from (11), the mean value theorem, and Theorem 1.2 that

$$\frac{d}{dt} F_t \bigg|_{t=s}$$

exists for $s > 0$ and has Fourier coefficients $(Da_n(s))^\infty_{-\infty}$. Then (12) gives (6). Finally,

$$|D^2a_n(t)| \le |n^2b_n| + |nc_n|, \qquad \text{all } n.$$

It follows that

$$\left(\frac{d}{dt}\right)^2 F_t\Bigg|_{t=s}$$

exists, $s > 0$. The choice of $a_n(t)$ implies that (4) holds. □

Let us look more closely at the distribution F_t in the case when G and H are the distributions defined by functions g and h in $\mathscr{P}$. First, suppose $h = 0$. The Fourier series for g converges:

$$g(x) = \sum_{-\infty}^{\infty} a_n \exp(inx).$$

Inequality (9) implies that the Fourier series for F_t also converges; then F_t is the distribution defined by the function $u_t \in \mathscr{P}$, where

$$u(x, t) = u_t(x) = \sum_{-\infty}^{\infty} a_n \cos nt \cdot \exp(inx)$$

$$= \frac{1}{2} \sum_{-\infty}^{\infty} a_n[\exp(int) + \exp(-int)] \exp(inx)$$

$$= \frac{1}{2} \sum_{-\infty}^{\infty} a_n \exp(in(x + t)) + \frac{1}{2} \sum_{-\infty}^{\infty} a_n \exp(in(x - t)),$$

or

(13) $$u(x, t) = \tfrac{1}{2}g(x + t) + \tfrac{1}{2}g(x - t).$$

It is easily checked that u is a solution of

$$\frac{\partial^2 u}{\partial t^2} = \frac{\partial^2 u}{\partial x^2}, \qquad u(x, 0) = g(x), \qquad \frac{\partial u}{\partial t}(x, 0) = 0.$$

Next, consider the case $g = 0$. Again $(a_n(t))^\infty_{-\infty}$ is the sequence of Fourier coefficients of a function $u_t \in P$,

$$u(x, t) = u_t(x) = \sum_{n \ne 0} n^{-1}c_n \sin nt \cdot \exp(inx) + c_0 t.$$

To get rid of the n^{-1} factor, we differentiate:

$$\frac{\partial}{\partial x} u(x, t) = \sum_{-\infty}^{\infty} ic_n \sin nt \exp(inx)$$

$$= \frac{1}{2} \sum_{-\infty}^{\infty} c_n[\exp(int) - \exp(-int)] \exp(inx)$$

$$= \frac{1}{2} \sum_{-\infty}^{\infty} c_n \exp(in(x + t)) - \frac{1}{2} \sum_{-\infty}^{\infty} c_n \exp(in(x - t))$$

$$= \tfrac{1}{2}h(x + t) - \tfrac{1}{2}h(x - t).$$

Integrating with respect to x gives

(13)'
$$u(x, t) = \frac{1}{2} \int_{x-t}^{x+t} h(y)\, dy + a(t)$$

where $a(t)$ is a constant to be determined so that u is periodic as a function of x. But the periodicity of h implies that $a(t)$ should be taken to be zero. Then u so defined is a solution of

$$\frac{\partial^2 u}{\partial t^2} = \frac{\partial^2 u}{\partial x^2}, \qquad u(x, 0) = 0, \qquad \frac{\partial u}{\partial t}(x, 0) = h(x).$$

Equation (1) for the vibrating string may be derived as follows. Approximate the curve u_t representing the displacement at time t by a polygonal line joining the points $\ldots(x - \varepsilon, u(x - \varepsilon, t)), (x, u(x, t)), (x + \varepsilon, u(x + \varepsilon, t)), \ldots$. The force on the string at the point x is due to the tension of the string. In this approximation the force due to tension is directed along the line segments from $(x, u(x, t))$ to $(x \pm \varepsilon, u(x \pm \varepsilon, t))$. The net *vertical* component of force is then proportional to

$$\left(\frac{\partial}{\partial t}\right)^2 u(x, t) \approx c^2 \varepsilon^{-2}[u(x + \varepsilon) + u(x - \varepsilon) - 2u(x)]$$

where c is constant. We take the limit as $\varepsilon \to 0$. Two applications of L'Hôpital's rule to the right side considered as a function of ε give (1).

The constant c^2 can be seen to be equal to

$$\kappa \tau r^{-2},$$

where r is the diameter of the string or wire, τ is the tension, and κ is proportional to the density of the material. Let us suppose also that the length of the string is πl instead of π. The solution of the problem corresponding to (1), (2), (3) when g is real and $h = 0$ can be shown to be of the form

(14)
$$u(x, t) = \sum_{n=1}^{\infty} b_n \cos(cnt/l)\sin(nx/l).$$

A single term

$$b_n \cos(cnt/l)\sin(nx/l)$$

represents a "standing wave", a sine curve with n maxima and minima in the interval $[0, \pi l]$ and with height varying with time according to the term $\cos(cnt/l)$. Thus the maximum height is b_n, and the original wave is repeated after a time interval of length

$$2\pi l/cn.$$

Thus the *frequency* for this term is

$$cn/2\pi l = n(\kappa\tau)^{1/2}/2\pi lr,$$

an integral multiple of the lowest frequency

(15)
$$(\kappa\tau)^{1/2}/2\pi lr.$$

In hearing the response of a plucked string the ear performs a Fourier analysis on the air vibrations corresponding to $u(x, t)$. Only finitely many terms of (14) represent frequencies low enough to be heard, so the series (14) is heard as though it were a finite sum

(16) $$\sum_{n=1}^{N} b_n \cos{(cnt/l)} \sin{(nx/l)}.$$

In general, if the basic frequency (15) is not too low (or high) it is heard as the *pitch* of the string, and the coefficients $b_1, b_2, \ldots, b_N$ determine the *purity*: a pure tone corresponds to all but one b_n being zero. Formula (15) shows that the pitch varies inversely with length and radius, and directly with the square root of the tension.

Exercises

1. In Theorem 5.1, suppose DG and H are in $\mathbf{L}^2$. Show that

$$DF_t \quad \text{and} \quad \frac{d}{dt} F_t$$

are in $\mathbf{L}^2$ for each t, and

$$\| DF_t \|^2 + \left\| \frac{d}{dt} F_t \right\|^2 = \| DG \|^2 + \| H \|^2.$$

2. In the problem (1), (2), (3) with $c = 1$ suppose g, h, and u are smooth functions. Show, by computing the derivative with respect to t, that

$$\int_0^\pi \left| \frac{\partial}{\partial x} u(x, t) \right|^2 dx + \int_0^\pi \left| \frac{\partial}{\partial t} u(x, t) \right|^2 dx$$

is constant. (This expresses *conservation of energy*: the first term represents potential energy, from the tension, while the second term represents kinetic energy.)

3. Use Exercise 2 and the results of this section to prove a theorem about existence and uniqueness of classical solutions of the problem (1), (2), (3).

4. Show that if $H(f) = 0$ when f is constant, then the solution $(F_t)_{t > 0}$ in Theorem 5.1 can be written in the form

$$F_t = \tfrac{1}{2}(T_t G + T_{-t} G) + \tfrac{1}{2}(T_{-t} SH - T_t SH).$$

Here again T_t denotes translation, while S is the operator from $\mathscr{P}'$ to $\mathscr{P}'$ defined by

$$SH(u) = H(v), \qquad u \in \mathscr{P},$$

where v is the periodic extension of the function

$$v_1(x) = \int_x^{2\pi} u(t) \, dt + \frac{x}{2\pi} \int_0^{2\pi} u(t) \, dt, \qquad 0 \le x \le 2\pi.$$

This is the analog for distributions of formulas (13) and (13)′.

§6. Laplace's equation and the Dirichlet problem

A third equation of mathematical and physical importance is *Laplace's equation*. In two variables this is the equation

(1)
$$\frac{\partial^2 u}{\partial x^2} + \frac{\partial^2 u}{\partial y^2} = 0.$$

A function u satisfying this equation is said to be *harmonic*. A typical problem connected with this equation is the *Dirichlet problem* for the disc: find a function continuous on the closed unit disc

$$\{(x, y) \mid x^2 + y^2 \le 1\}$$

and equal on the boundary to a given function g:

(2)
$$u(x, y) = g(x, y), \qquad x^2 + y^2 = 1.$$

A physical situation leading to this problem is the following. Let $u(x, y, t)$ denote the temperature at time t at the point with coordinates (x, y) on a metal disc of radius 1. Suppose the temperature at the edge of the disc is fixed, though varying from point to point, while the interior of the disc is insulated. Eventually thermal equilibrium will be reached: u will be independent of time. The resulting temperature distribution u is approximately a solution of (1), (2).

To solve this equation we express u and g by polar coordinates:

$$u = u(r, \theta), \qquad g = g(\theta)$$

where

$$x = r \cos \theta, \qquad y = r \sin \theta; \qquad r = (x^2 + y^2)^{1/2}, \qquad \theta = \tan^{-1}(yx^{-1}).$$

Then

$$\frac{\partial u}{\partial x} = \frac{\partial u}{\partial r}\frac{\partial r}{\partial x} + \frac{\partial u}{\partial \theta}\frac{\partial \theta}{\partial x} = \cos\theta\frac{\partial u}{\partial r} - \frac{\sin\theta}{r}\frac{\partial u}{\partial \theta},$$

and similarly for $\partial u/\partial y$. An elementary but tedious calculation gives

$$\frac{\partial^2 u}{\partial x^2} + \frac{\partial^2 u}{\partial y^2} = \frac{\partial^2 u}{\partial r^2} + \frac{1}{r}\frac{\partial u}{\partial r} + \frac{1}{r^2}\frac{\partial^2 u}{\partial \theta^2}.$$

Thus we want to solve

(3)
$$r^2\frac{\partial^2 u}{\partial r^2} + r\frac{\partial u}{\partial r} + \frac{\partial^2 u}{\partial \theta^2} = 0, \qquad 0 \le r < 1, \qquad \theta \in \mathbb{R},$$

(4)
$$u(1, \theta) = g(\theta), \qquad \theta \in \mathbb{R},$$

where u and g are periodic in θ. We proceed formally. Suppose g has Fourier coefficients $(b_n)_{-\infty}^{\infty}$, and $u_r(\theta) = u(r, \theta)$ has Fourier coefficients

$$(a_n(r))_{-\infty}^{\infty}, \qquad 0 < r < 1.$$

Then (3) leads to the equation

(5) $$r^2 D^2 a_n + r D a_n - n^2 a_n = 0, \qquad 0 < r < 1.$$

From (4) we get

(6) $$a_n(1) = b_n,$$

and since $u_0(\theta)$ is constant we want

(7) $$a_n(0) = 0, \qquad n \neq 0.$$

We look for a solution $a_n(r)$ of the form $b_n r^c$, where $c = c(n)$ is a constant, for each n. This will be a solution if and only if

$$c(c - 1) + c - n^2 = 0,$$

or $c^2 = n^2$. Then (7) gives $c = |n|$. We are led to the formal solution

$$u(r, \theta) = \sum_{-\infty}^{\infty} b_n r^{|n|} \exp(in\theta).$$

Formally, this should be the convolution of g with the distribution whose Fourier coefficients are

$$(r^{|n|})_{-\infty}^{\infty}.$$

For $r < 1$ these are the Fourier coefficients of a function $P_r \in \mathscr{P}$. In fact,

(8) $$P_r(\theta) = \sum_{-\infty}^{\infty} r^{|n|} e^{in\theta}$$

$$= 1 + \sum_{1}^{\infty} (re^{i\theta})^n + \sum_{1}^{\infty} (re^{-i\theta})^n$$

$$= (1 - r^2)|1 - re^{i\theta}|^{-2} = (1 - r^2)(1 - 2r\cos\theta + r^2)^{-1}.$$

Note that $P_r(\theta)$ is an approximate identity as $r \to 1$: the first expression on the right in (8) shows

$$\frac{1}{2\pi} \int_0^{2\pi} P_r(\theta) \, d\theta = 1$$

and the last expression shows that $P_r(\theta) \geq 0$ and

$$\lim_{r \to 1} \int_\delta^{2\pi - \delta} P_r(\theta) \, d\theta = 0$$

for each $0 < \delta < \pi$. The function

$$P(r, \theta) = P_r(\theta)$$

is called the *Poisson kernel* for the Dirichlet problem in the unit disc.

Theorem 6.1. *Suppose F is a periodic distribution. There is a unique function $u(r, \theta)$ defined and smooth in the open unit disc, satisfying (3), and such that the distributions defined by the functions $u_r(\theta) = u(r, \theta)$ converge to F in the sense of $\mathcal{P}'$ as $r \to 1$. Moreover, if $F = F_g$ where $g \in \mathcal{C}$, then $u_r \to g$ uniformly as $r \to 1$. The functions u_r are given by convolution with the Poisson kernel:*

$$u_r = F * P_r.$$

Proof. Uniqueness was proved in the derivation above. Let $u_r = F * P_r$. Since P_r is an approximate identity, we do have $u_r \to F$ ($\mathcal{P}'$) as $r \to 1$, and $u_r \to g$ uniformly if $F = F_g$, $g \in \mathcal{C}$. We must show that u is smooth and satisfies (3). Note that when $F = F_g$, $g \in \mathcal{C}$, then explicitly

$$u(r, \theta) = \frac{1}{2\pi} \int_0^{2\pi} P(r, \theta - \varphi) g(\varphi) \, d\varphi$$

and we may differentiate under the integral sign to prove that u is smooth. Moreover, since $P(r, \theta)$ satisfies (3), so does u.

Finally, suppose u has merely a distribution F as its value on the boundary. Note that if $0 \le r, s < 1$ then (by computing Fourier coefficients, for example)

$$P_r * P_s = P_{rs}.$$

In particular, choose any $R > 0$, $R < 1$. It suffices to show that u is smooth in the disc $r < R$ and satisfies (3) there. But when $r < R$,

$$s = rR^{-1} < 1$$

and

$$u_r = F * P_R = F * (P_R * P_s) = (F * P_R) * P_s = u_R * P_s.$$

Since $P_R \in \mathcal{P}$, u_R is a smooth function of θ. Then

$$u(r, \theta) = \frac{1}{2\pi} \int_0^{2\pi} P(rR^{-1}, \theta - \varphi) u_R(\varphi) \, d\varphi,$$

$0 \le r < R$. Again, differentiation under the integral sign shows that u is smooth and satisfies (3).   □

The preceding theorem leads to the remarkable result that a real-valued harmonic function is (locally) the real part of a function defined by a convergent power series in $z = x + iy$.

Theorem 6.2. *Suppose u is a harmonic real-valued function of class C^2 defined in an open subset of $\mathbb{R}^2$ containing the point (x_0, y_0). Then there is a function f defined by a convergent power series:*

$$f(z) = \sum_0^\infty a_n(z - z_0)^n, \qquad |z - z_0| < \varepsilon, \qquad z_0 = x_0 + iy_0,$$

such that

$$u(x, y) = \operatorname{Re} f(x + iy)$$

when

$$|x + iy - z_0| < \varepsilon.$$

Proof. Suppose first that $(x_0, y_0) = (0, 0)$ and that the set in which u is defined contains the closed disc $x^2 + y^2 \le 1$. Let

$$g(\theta) = u(\cos \theta, \sin \theta), \qquad \theta \in \mathbb{R}.$$

Then u is the unique solution of the Dirichlet problem in the unit disc with g as value on the boundary. If $(b_n)_{-\infty}^{\infty}$ are the Fourier coefficients of g, then we know that in polar coordinates u is given by

$$\sum_{-\infty}^{\infty} r^{|n|} b_n \exp{(in\theta)}.$$

Since u is real, g is real. Therefore $b_n = b_{-n}^*$ and the series is

$$b_0 + 2 \operatorname{Re} \left(\sum_{1}^{\infty} r^n b_n \exp{(in\theta)} \right).$$

Let f be defined by

$$f(z) = \sum_{0}^{\infty} a_n z^n$$

where

$$a_0 = b_0; \qquad a_n = 2b_n, \qquad n > 0.$$

Then

$$u(x, y) = \operatorname{Re} \left(\sum_{0}^{\infty} a_n (re^{i\theta})^n \right) = \operatorname{Re}{(f(re^{i\theta}))}$$

$$= \operatorname{Re}{(f(x + iy))}, \qquad x^2 + y^2 < 1.$$

In the general case, assume that u is defined on a set containing the closed disc of radius ε centered at (x_0, y_0), and let

$$u_1(x, y) = u(x_0 + \varepsilon x, y_0 + \varepsilon y).$$

Then u_1 is harmonic in a set containing the unit disc, so

$$u_1 = \operatorname{Re} f_1,$$

f_1 defined by a power series in the unit disc. Then

$$u = \operatorname{Re} f, \qquad (x - x_0)^2 + (y - y_0)^2 < \varepsilon,$$

where

$$f(x, y) = f_1(\varepsilon^{-1}(x - x_0), \varepsilon^{-1}(y - y_0))$$

is defined by a power series in the disc of radius ε around $z_0 = x_0 + iy_0$. $\quad\square$

Exercises

1. Prove the converse of Theorem 6.2: if

$$f(z) = \sum_{0}^{\infty} a_n(z - z_0)^n, \qquad |z - z_0| < R,$$

and we let

$$u(x, y) = \operatorname{Re} f(x + iy), \qquad |x + iy - z_0| < R,$$

then u is harmonic.

2. There is a *maximum principle* for harmonic functions analogous to the maximum principle for solutions of the heat equation discussed in §4.

(a) Show that if u is of class C^2 on an open set A in $\mathbb{R}^2$ and

$$\frac{\partial^2 u}{\partial x^2} + \frac{\partial^2 u}{\partial y^2} > 0$$

at each point of A, then u does not have a local maximum at any point of A.

(b) Suppose u is of class C^2 and harmonic in an open disc in $\mathbb{R}^2$ and continuous on the closure of this disc. Show, by considering the functions

$$u_\varepsilon(x, y) = u(x, y) + \varepsilon x^2 + \varepsilon y^2$$

that u attains its maximum on the boundary of the disc.

3. Use the result of Exercise 2 to give a second proof of the uniqueness of the solution of the Dirichlet problem for a continuous boundary function g.

4. Suppose u is continuous on the closed disc $x^2 + y^2 \le R$ and harmonic in the open disc $x^2 + y^2 < R$. Give a formula for $u(x, y)$ (or $u(r, \theta)$) for $x^2 + y^2 < R$ in terms of the values of u for $x^2 + y^2 = R$. Give formulas for the derivatives $\partial u/\partial r$ and $\partial u/\partial \theta$ also.

5. Suppose u is defined on all of $\mathbb{R}^2$ and is harmonic. Use the result of Exercise 4 to show that if u is bounded, then it is constant.

Chapter 6

Complex Analysis

§1. Complex differentiation

Suppose Ω is an open subset of the complex plane $\mathbb{C}$. Recall that this means that for each $z_0 \in \Omega$ there is a $\delta > 0$ so that Ω contains the disc of radius δ around z_0:

$$z \in \Omega \qquad \text{if } |z - z_0| < \delta.$$

A function $f: \Omega \to \mathbb{C}$ is said to be *differentiable* at $z \in \Omega$ if the limit

$$\lim_{w \to z} \frac{f(w) - f(z)}{w - z}$$

exists. If so, the limit is called the derivative of f at z and denoted $f'(z)$.

These definitions are formally the same as those given for functions defined on open subsets of $\mathbb{R}$, and the proofs of the three propositions below are also identical to the proofs for functions of a real variable.

Proposition 1.1. *If $f: \Omega \to \mathbb{C}$ is differentiable at $z \in \Omega$, then it is continuous at z.*

Proposition 1.2. *Suppose $f: \Omega \to \mathbb{C}$, $g: \Omega \to \mathbb{C}$ and $a \in \mathbb{C}$. If f and g are differentiable at $z = \Omega$, then so are the functions af, $f + g$, and fg:*

$$(af)'(z) = af'(z)$$
$$(f + g)'(z) = f'(z) + g'(z)$$
$$(fg)'(z) = f'(z)g(z) + f(z)g'(z).$$

If also $g(z) \neq 0$, then f/g is differentiable at z and

$$(f/g)'(z) = [f'(z)g(z) - f(z)g'(z)]g(z)^{-2}.$$

Proposition 1.3 (Chain rule). *Suppose f is differentiable at $z \in \mathbb{C}$ and g is differentiable at $f(z)$. Then the compositive function $g \circ f$ is differentiable at z and*

$$(g \circ f)'(z) = g'(f(z))f'(z).$$

The proof of the following theorem is also identical to the proof of the corresponding Theorem 4.4 of Chapter 2.

Theorem 1.4. *Suppose f is defined by a convergent power series:*

$$f(z) = \sum_{n=0}^{\infty} a_n(z - z_0)^n, \qquad |z - z_0| < R.$$

Then f is differentiable at each point z with $|z - z_0| < R$, and

$$f'(z) = \sum_{n=1}^{\infty} na_n(z - z_0)^{n-1}.$$

In particular, the exponential and the sine and cosine functions are differentiable as functions defined on $\mathbb{C}$, and $(\exp z)' = \exp z$, $(\cos z)' = -\sin z$, $(\sin z)' = \cos z$.

A remarkable fact about complex differentiation is that a converse of Theorem 1.4 is true: if f is defined in the disc $|z - z_0| < R$ and differentiable at each point of this disc, then f can be expressed as the sum of a power series which converges in the disc. We shall sketch one proof of this fact in the Exercises at the end of this section, and give a second proof in §3 and a third in §7 (under the additional hypothesis that the derivative is continuous). Here we want to give some indication why the hypothesis of differentiability is so much more powerful in the complex case than in the real case. Consider the function

$$f(z) = z^*, \quad \text{or} \quad f(x + iy) = x - iy.$$

Take $t \in \mathbb{R}$, $t \neq 0$. Then

$$t^{-1}[f(z + t) - f(z)] = 1,$$
$$(it)^{-1}[f(z + it) - f(z)] = -1,$$

so the ratio

$$w^{-1}[f(w) - f(z)]$$

depends on the direction of the line through w and z, even in the limit as $w \to z$. Therefore this function f is not differentiable at any point.

Given $f: \Omega \to \mathbb{C}$, define functions u, v by

$$u(x, y) = \operatorname{Re} f(x + iy) = \tfrac{1}{2}f(x + iy) + \tfrac{1}{2}(f(x + iy))^*,$$

$$v(x, y) = \operatorname{Im} f(x + iy) = \frac{i}{2}(f(x + iy))^* - \frac{i}{2}f(x + iy).$$

Thus

$$f(x + iy) = u(x, y) + iv(x, y).$$

We shall speak of u and v as the *real* and *imaginary* parts of f and write

$$f = u + iv.$$

(This is slightly incorrect, since f is being considered as a function of $z \in \Omega \subset \mathbb{C}$, while u and v are considered as functions of two *real* variables x, y.)

Theorem 1.5. *Suppose $\Omega \subset \mathbb{C}$ is open, $f: \Omega \to \mathbb{C}$, $f = u + iv$. Then if f is differentiable at $z = x + iy \in \Omega$, the partial derivatives*

(1) $\dfrac{\partial u}{\partial x}, \quad \dfrac{\partial u}{\partial y}, \quad \dfrac{\partial v}{\partial x}, \quad \dfrac{\partial v}{\partial y}$

all exist at (x, y) *and satisfy*

(2)
$$\frac{\partial}{\partial x} u(x, y) = \frac{\partial}{\partial y} v(x, y)$$

(3)
$$\frac{\partial}{\partial y} u(x, y) = -\frac{\partial}{\partial x} v(x, y).$$

Conversely, suppose that the partial derivatives (1) *all exist and are continuous in an open set containing* (x, y) *and satisfy* (2), (3) *at* (x, y). *Then f is differentiable at* $z = x + iy$.

The equations (2) and (3) are called the *Cauchy-Riemann equations*. They provide a precise analytical version of the requirement that the limit defining $f'(z)$ be independent of the direction of approach. Note that in the example $f(z) = z^*$ we have

$$\frac{\partial u}{\partial x} = 1, \qquad \frac{\partial v}{\partial y} = -1, \qquad \frac{\partial u}{\partial y} = \frac{\partial v}{\partial x} = 0.$$

Proof. Suppose f is differentiable at $z = x + iy$. Then

(4) $$\lim_{t \to 0} t^{-1}[f(z + t) - f(z)] = f'(z) = \lim_{t \to 0} (it)^{-1}[f(z + it) - f(z)].$$

The left side of (4) is clearly

$$\frac{\partial}{\partial x} u(x, y) + i \frac{\partial}{\partial x} v(x, y),$$

while the right side is

$$-i \frac{\partial}{\partial y} u(x, y) + \frac{\partial}{\partial y} v(x, y).$$

Equating the real and imaginary parts of these two expressions, we get (2) and (3).

Conversely, suppose the first partial derivatives of u and v exist and are continuous near (x, y), and suppose (2) and (3) are true. Let

$$h = a + ib$$

where a and b are real and near zero, $h \neq 0$. We apply the Mean Value Theorem to u and v to get

$$f(z + h) - f(z) = f(z + a + ib) - f(z + a) + f(z + a) - f(z)$$

$$= \frac{\partial}{\partial y} u(x + a, y + t_1 b)b + i \frac{\partial}{\partial y} v(x + a, y + t_2 b)b$$

$$+ \frac{\partial}{\partial x} u(x + t_3 a, y)a + i \frac{\partial}{\partial y} v(x + t_4 a, y)a,$$

where $0 < t_j < 1, j = 1, 2, 3, 4$. Because of (2) and (3),

$$\left[\frac{\partial}{\partial x} u(x, y) + i \frac{\partial}{\partial y} v(x, y) \right] h$$

$$= \frac{\partial}{\partial y} u(x, y)b + i \frac{\partial}{\partial y} v(x, y)b + \frac{\partial}{\partial x} u(x, y)a + i \frac{\partial}{\partial x} v(x, y)a.$$

Therefore

$$h^{-1}[f(z + h) - f(z)] - \left[\frac{\partial}{\partial x} u(x, y) + i \frac{\partial}{\partial y} v(x, y) \right]$$

is a sum of four terms similar in size to

$$\frac{\partial}{\partial y} u(x + a, y + t_1 b) - \frac{\partial}{\partial y} u(x, y).$$

Since the partial derivatives were assumed continuous, these terms $\to 0$ as $h \to 0$. $\square$

A function $f: \Omega \to \mathbb{C}$, Ω open in $\mathbb{C}$, is said to be *holomorphic* in Ω if it is differentiable at each point of Ω and the derivative f' is a continuous function. (Actually, the derivative is *necessarily* continuous if it exists at each point; later we shall indicate how this may be proved.) Theorem 1.5 has the following immediate consequence.

Corollary 1.6. $f: \Omega \to \mathbb{C}$ *is holomorphic in* Ω *if and only if its real and imaginary parts*, u *and* v, *are of class* C^1 *and satisfy the Cauchy-Riemann equations* (2) *and* (3) *at each point* (x, y) *such that* $x + iy \in \Omega$.

Locally, at least, a holomorphic function can be integrated.

Corollary 1.7. *Suppose* g *is holomorphic in a disc* $|z - z_0| < R$. *Then there is a function* f, *holomorphic for* $|z - z_0| < R$, *such that* $f' = g$.

Proof. Let u, v be the real and imaginary parts of g. We want to determine real functions q, r such that

$$f = q + ir$$

has derivative g. Because of Theorem 1.5 we can see that this will be true if and only if

$$\frac{\partial q}{\partial x} = \frac{\partial r}{\partial y} = u, \qquad \frac{\partial q}{\partial y} = -\frac{\partial r}{\partial x} = -v.$$

The condition that u, $-v$ be the partial derivatives of a function q is (by Exercises 1 and 2 of §7, Chapter 2)

$$\frac{\partial u}{\partial y} = -\frac{\partial v}{\partial x}.$$

The condition that v and u be the partial derivatives of a function r is

$$\frac{\partial v}{\partial y} = \frac{\partial u}{\partial x}.$$

Thus there are functions q, r with the desired properties.   ☐

Exercises

1. Let $f(x + iy) = x^2 + y^2$. Show that f is differentiable only at $z = 0$.

2. Suppose $f: \mathbb{C} \to \mathbb{R}$. Show that f is differentiable at every point if and only if f is constant.

3. Let $f(0) = 0$ and

$$f(x + iy) = 2xy(x^2 + y^2)^{-1}, \qquad x + iy \neq 0.$$

Show that the first partial derivatives of f exist at each point and are both zero at $x = y = 0$. Show that f is not differentiable (in fact not continuous) at $z = 0$. Why does this not contradict Theorem 1.5?

4. Suppose f is holomorphic in Ω and suppose the real and imaginary parts u, v are of class C^2 in Ω. Show that u and v are harmonic.

5. Suppose f is as in Exercise 4, and suppose the disc $|z - z_0| \leq R$ is contained in Ω. Use Exercise 4 together with Theorem 6.2 of Chapter 5 to show that there is a power series $\sum a_n(z - z_0)^n$ converging to $f(z)$ for $|z - z_0| < R$.

6. Suppose g is holomorphic in Ω, and suppose Ω contains the disc $|z - z_0| \leq R$. Let f be such that $f'(z) = g(z)$ for $|z - z_0| \leq R$ (using Corollary 1.7). Show that the real and imaginary parts of f are of class C^2.

7. Use the results of Exercises 5 and 6, together with Theorem 1.4, to prove the following theorem.

If g is holomorphic in Ω and $z_0 \in \Omega$, then there is a power series such that

$$g(z) = \sum_{n=0}^{\infty} a_n(z - z_0)^n, \qquad |z - z_0| < R,$$

for any R such that Ω contains the disc of radius R with center z_0.

§2. Complex integration

Suppose $\Omega \subset \mathbb{C}$ is open. A *curve* in Ω is, by definition, a continuous function γ from a closed interval $[a, b] \subset \mathbb{R}$ into Ω. The curve γ is said to be *smooth* if it is a function of class C^1 on the open interval (a, b) and if the one-sided derivatives exist at the endpoints:

$$(t - a)^{-1}[\gamma(t) - \gamma(a)] \quad \text{converges as} \quad t \to a, t > a;$$
$$(t - b)^{-1}[\gamma(t) - \gamma(b)] \quad \text{converges as} \quad t \to b, t < b.$$

The curve γ is said to be *piecewise smooth* if there are points $a_0, a_1, \ldots, a_r$ with

$$a = a_0 < a_1 < \cdots < a_r = b$$

such that the restriction of γ to $[a_{j-1}, a_j]$ is a smooth curve, $1 \leq j \leq r$. An example is

$$\gamma(t) = z_0 + \varepsilon \exp(it), \qquad t \in [0, 2\pi];$$

then the image

$$\{\gamma(t) \mid t \in [0, 2\pi]\}$$

is the circle of radius ε around z_0. This is a smooth curve. A second example is

$$\begin{aligned}
\gamma(t) &= t, & t &\in [0, 1], \\
\gamma(t) &= 1 + i(t - 1), & t &\in (1, 2], \\
\gamma(t) &= 1 + i - (t - 2), & t &\in (2, 3], \\
\gamma(t) &= i - (t - 3)i, & t &\in (3, 4].
\end{aligned}$$

Here γ is piecewise smooth and the image is a unit square.

Suppose $\gamma \colon [a, b] \to \Omega$ is a curve and $f \colon \Omega \to \mathbb{C}$. The integral of f over γ,

$$\int_\gamma f,$$

is defined to be the limit, as the mesh of the partition $P = (t_0, t_1, \ldots, t_n)$ of $[a, b]$ goes to zero, of

$$\sum_{j=1}^{n} f(\gamma(t_i))[\gamma(t_i) - \gamma(t_{i-1})].$$

Proposition 2.1. *If γ is a piecewise smooth curve in Ω and $f \colon \Omega \to \mathbb{C}$ is continuous, then the integral of f over γ exists and*

$$(1) \qquad \int_\gamma f = \int_a^b f(\gamma(t))\gamma'(t)\, dt.$$

Proof. The integral on the right exists, since the integrand is bounded and is continuous except possibly at finitely many points of $[a, b]$. To prove that (1) holds we assume first that γ is smooth. Let $P = (t_0, t_1, \ldots, t_n)$ be a partition of $[a, b]$. Then

$$(2) \qquad \sum f(\gamma(t_i))[\gamma(t_i) - \gamma(t_{i-1})] = \sum f(\gamma(t_i))\gamma'(t_i)[t_i - t_{i-1}] + R,$$

where

$$|R| \leq \sup |f(\gamma(t))| \cdot \sum |\gamma(t_i) - \gamma(t_{i-1}) - \gamma'(t_i)(t_i - t_{i-1})|.$$

Applying the Mean Value Theorem to the real and imaginary parts of γ on $[t_{i-1}, t_i]$ and using the continuity of γ', we see that

$$R \to 0 \quad \text{as the mesh} \quad |P| \to 0.$$

On the other hand, the sum on the right side of (2) is a Riemann sum for the integral on the right side of (1). Thus we have shown that the limit exists and (1) is true.

If γ is only piecewise smooth, then the argument above breaks down on intervals containing points of discontinuity of γ'. However, the total contribution to the sums in (2) from such intervals is easily seen to be bounded in modulus by a constant times the mesh of P. Thus again the limit exists as $|P| \to 0$ and (1) is true. □

Note that the integral $\int_\gamma f$ depends not only on the set of points

$$\{\gamma(t) \mid t \in [a, b]\}$$

but also on the "sense," or ordering, of them. For example, if

$$\gamma_1(t) = \exp(it), \qquad \gamma_2(t) = \exp(-it), \qquad t \in [0, 2\pi],$$

then the point sets are the same but

$$\int_{\gamma_1} f = -\int_{\gamma_2} f.$$

Furthermore, it matters how many times the point set is traced out by γ: if

$$\gamma_3(t) = \exp(int),$$

then

$$\int_{\gamma_3} f = n \int_{\gamma_1} f.$$

A curve $\gamma: [a, b] \to \Omega$ is said to be *constant* if γ is a constant function. If so, then

$$\int_\gamma f = 0, \qquad \text{all } f.$$

A curve $\gamma: [a, b] \to \Omega$ is said to be *closed* if $\gamma(a) = \gamma(b)$. (All the examples given so far have been examples of closed curves.) Two closed curves $\gamma_0, \gamma_1: [a, b] \to \Omega$ are said to be *homotopic in* Ω if there is a continuous function

$$\Gamma: [a, b] \times [0, 1] \to \Omega$$

such that

$$\Gamma(t, 0) = \gamma_0(t), \qquad \Gamma(t, 1) = \gamma_1(t), \qquad \text{all } t \in [a, b],$$
$$\Gamma(a, s) = \Gamma(b, s), \qquad \text{all } s \in [0, 1].$$

The function Γ is called a *homotopy* from γ_0 to γ_1. If Γ is such a homotopy, let

$$\gamma_s(t) = \Gamma(t, s), \qquad s \in (0, 1).$$

Then each γ_s is a closed curve, and we think of these as being a family of curves varying continuously from γ_0 to γ_1, within Ω.

Theorem 2.2 (Cauchy's Theorem). *Suppose* $\Omega \subset \mathbb{C}$ *is open and suppose* f *is holomorphic in* Ω. *Suppose* γ_0 *and* γ_1 *are two piecewise smooth closed curves in* Ω *which are homotopic in* Ω. *Then*

$$\int_{\gamma_0} f = \int_{\gamma_1} f.$$

The importance of this theorem can scarcely be overestimated. We shall first cite a special case of the theorem as a Corollary, and prove the special case.

Corollary 2.3. *Suppose Ω is either a disc*

$$\{z \mid |z - z_0| < R\}$$

or a rectangle

$$\{x + iy \mid x_1 < x < x_2, y_1 < y < y_2\}.$$

Suppose f is holomorphic in Ω and γ is any piecewise smooth closed curve in Ω. Then

$$\int_\gamma f = 0.$$

Proof. It is easy to see that in this case γ is homotopic to a constant curve γ_0, so that the conclusion follows from Cauchy's Theorem. However, let us give a different proof. By Corollary 1.7, or by the analogous result for a rectangle in place of a disc, there is a function h, holomorphic in Ω, such that $h' = f$. But then

$$\int_\gamma f = \int_a^b f(\gamma(t))\gamma'(t)\, dt = \int_a^b h'(\gamma(t))\gamma'(t)\, dt$$

$$= \int_a^b [h \circ \gamma]'(t)\, dt = h(\gamma(b)) - h(\gamma(a)) = 0,$$

since $\gamma(a) = \gamma(b)$. □

Proof of Cauchy's Theorem. Let Γ be a homotopy from γ_0 to γ_1 and let

$$\gamma_s(t) = \Gamma(t, s), \qquad t \in [a, b], \qquad s \in (0, 1).$$

Assume for the moment that each curve γ_s is piecewise smooth. We would like to show that the integral of f over γ_s is independent of s, $0 \le s \le 1$.

Assume first that Γ is of class C^2 on the square $[a, b] \times (0, 1)$, that the first partial derivatives are uniformly bounded, and that

$$\gamma_s' \to \gamma_0' \quad \text{as} \quad s \to 0,$$
$$\gamma_s' \to \gamma_1' \quad \text{as} \quad s \to 1$$

uniformly on each interval $(c, d) \subset [a, b]$ on which γ_0' or γ_1', respectively, is continuous. These assumptions clearly imply that

$$F(s) = \int_{\gamma_s} f = \int_a^b f(\Gamma(t, s)) \frac{\partial \Gamma}{\partial t}(t, s)\, dt$$

is a continuous function of s, $s \in [0, 1]$. Furthermore under these assumptions we may apply results of §7 of Chapter 2 and differentiate under the integral sign when $s \in (0, 1)$ to get

$$F'(s) = \int_a^b \frac{\partial}{\partial s} \left(f(\Gamma(t, s)) \frac{\partial}{\partial t} \Gamma(t, s) \right) dt$$

$$= \int_a^b f'(\Gamma(t, s)) \frac{\partial}{\partial s} \Gamma(t, s) \frac{\partial}{\partial t} \Gamma(t, s) \, dt$$

$$+ \int_a^b f(\Gamma(t, s)) \frac{\partial^2}{\partial t \, \partial s} \Gamma(t, s) \, dt$$

$$= \int_a^b \frac{\partial}{\partial t} \left[f(\Gamma(t, s)) \frac{\partial}{\partial s} \Gamma(t, s) \right] dt$$

$$= f(\Gamma(b, s)) \frac{\partial}{\partial s} \Gamma(b, s) - f(\Gamma(a, s)) \frac{\partial}{\partial s} \Gamma(a, s)$$

$$= 0$$

(since $\Gamma(b, s) = \Gamma(a, s)$). Thus $F(0) = F(1)$.

Finally, do not assume Γ is differentiable. We may extend Γ in a unique way so as to be periodic in the first variable with period $b - a$, and even and periodic in the second variable with period 2; then $\Gamma: \mathbb{R} \times \mathbb{R} \to \Omega$. Let $\varphi: \mathbb{R} \to \mathbb{R}$ be a smooth function such that

$$\varphi(x) \geq 0, \qquad \text{all } x,$$

$$\int \varphi(x) \, dx = 1$$

$$\varphi(x) = 0 \qquad \text{if } |x| \geq 1.$$

Let $\varphi_n(x) = n\varphi(nx)$, $n = 1, 2, 3, \ldots$. Then

$$\int \varphi_n(x) \, dx = 1, \qquad \varphi_n(x) = 0 \text{ if } |x| \geq n^{-1}.$$

Let

$$\Gamma_n(t, s) = \int \Gamma(t - x, s - y) \varphi_n(x) \varphi_n(y) \, dx \, dy,$$

$(t, s) \in \mathbb{R} \times \mathbb{R}$. Then (see the arguments in §3 of Chapter 3) Γ_n is also periodic with the same periods as Γ, and $\Gamma_n \to \Gamma$ uniformly as $n \to \infty$. It follows from this (and the fact that $\Gamma(\mathbb{R} \times \mathbb{R})$ is a compact subset of Ω) that $\Gamma_n(\mathbb{R} \times \mathbb{R}) \subset \Omega$ if n is sufficiently large. Furthermore, the argument above shows that

$$\int_{\gamma_{0,n}} f = \int_{\gamma_{1,n}} f,$$

where

$$\gamma_{s,n}(t) = \Gamma_n(t, s), \qquad t \in [a, b], \qquad s \in [0, 1].$$

All we need do to complete the proof is show that

(3) $\int_{\gamma_{s,n}} f \to \int_{\gamma_s} f$ as $n \to \infty$, $s = 0$ or 1.

Note that we may choose the homotopy Γ so that $\gamma_s = \gamma_0$ for $0 \leq s \leq \frac{1}{3}$ and $\gamma_s = \gamma_1$ for $\frac{2}{3} \leq s \leq 1$; to see this, let Γ_0 be any homotopy from γ_0 to γ_1 in Ω, and let

$$\begin{aligned}
\Gamma(t, s) &= \gamma_0(t), & 0 \leq s \leq \tfrac{1}{3}, \\
\Gamma(t, s) &= \Gamma_0(t, 3s - 1), & \tfrac{1}{3} < s < \tfrac{2}{3}, \\
\Gamma(t, s) &= \gamma_1(t), & \tfrac{2}{3} \leq s \leq 1.
\end{aligned}$$

If Γ has these properties, then when $n \geq 3$ we have

$$\Gamma_n(t, s) = \int \Gamma(t - x, s)\varphi_n(x)\, dx, \qquad s = 0 \text{ or } 1.$$

It follows that $\gamma_{s,n} \to \gamma_s$ uniformly, $s = 0$ or 1. It also follows, by differentiating with respect to t, that $\gamma'_{s,n}$ is uniformly bounded and

$$\gamma'_{s,n} \to \gamma'_s$$

on each interval of $[0, 1]$ where γ'_s is continuous, $s = 0$ or 1. Therefore (3) is true. □

Exercises

1. Suppose $\gamma_1 \colon [a, b] \to \Omega$ and $\gamma_2 \colon [c, d] \to \Omega$ are two piecewise smooth curves with the same image:

$$C = \{\gamma_1(t) \mid t \in [a, b]\} = \{\gamma_2(t) \mid t \in [c, d]\}.$$

Suppose these curves trace out the image in the *same direction*, i.e., if

$$s, t \in [a, b],\ s', t' \in [c, d]$$

and

$$\gamma_1(s) = \gamma_2(s') \neq \gamma_1(t) = \gamma_2(t')$$

and

$$s < t$$

then

$$s' < t'.$$

Show that for any continuous $f \colon \Omega \to \mathbb{C}$ (not necessarily holomorphic),

$$\int_{\gamma_1} f = \int_{\gamma_2} f.$$

This justifies writing the integral as an integral over the point set C:

$$\int_{\gamma_1} f = \int_C f(z)\, dz,$$

where we tacitly assume a direction chosen on C.

2. Suppose C in Exercise 1 is a circle of radius R and suppose $|f(z)| \le M$, all $z \in C$. Show that

$$\left| \int_C f(z)\, dz \right| \le M \cdot 2\pi R.$$

3. Let $\gamma(t) = z_0 + \varepsilon e^{it}$, $t \in [0, 2\pi]$, where $\varepsilon > 0$. Show that

$$\int_\gamma (z - z_0)^{-1} = 2\pi i.$$

4. Let $\gamma_n(t) = \exp(int)$, $t \in [0, 2\pi]$, $n = 0, \pm 1, \pm 2, \ldots$. Show that

$$\int_{\gamma_n} z^{-1} = 2n\pi i.$$

5. Let $\Omega = \{z \in \mathbb{C} \mid z \ne 0\}$. Use the result of Exercise 4 to show that the curves γ_n and γ_m are not homotopic in Ω if $n \ne m$. Show that each γ_n is homotopic to γ_0 in $\mathbb{C}$, however.

6. Use Exercise 4 to show that there is no function f, defined for all $z \ne 0$, such that $f'(z) = z^{-1}$, all $z \ne 0$. Compare this to Corollary 1.7.

7. Let Ω be a disc with a point removed:

$$\Omega = \{z \mid |z - z_0| < R,\ z \ne z_1\},$$

where $|z - z_1| < R$. Let γ_0 and γ_1 be two circles in Ω enclosing z_1, say

$$\begin{aligned}
\gamma_0(t) &= z_1 + \varepsilon e^{it}, & t &\in [0, 2\pi], \\
\gamma_1(t) &= z + r e^{it}, & t &\in [0, 2\pi].
\end{aligned}$$

Here $|z - z_1| < r < R$ and $\varepsilon > 0$ is chosen so that $|\gamma_1(t) - z_1| > \varepsilon$, all t. Construct a homotopy from γ_0 to γ_1.

8. Suppose Ω contains the square

$$\{x + iy \mid 0 \le x, y \le 1\}.$$

Suppose f is differentiable at each point of Ω; here we do *not* assume that f' is continuous. Let C be the boundary of the square, with the counterclockwise direction. Show that

$$\int_C f(z)\, dz = 0.$$

This extension of Corollary 2.3 is due to Goursat. (Hint: for each integer $k > 0$, divide the square into 4^k smaller squares with edges of length 2^{-k}.

Let $C_{k,1}, \ldots, C_{k,4^k}$ be the boundaries of these smaller squares, with the counterclockwise direction. Then show that

$$\int_C f(z)\,dz = \sum_j \int_{C_{k,j}} f(z)\,dz.$$

It follows that if

$$\left| \int_C f(z)\,dz \right| = M > 0,$$

then for each k there is a $j = j(k)$ such that

$$\left| \int_{C_k} f(z)\,dz \right| \geq 4^{-k}M,$$

where $C_k = C_{k,j}$. Let z_k be the center of the square with boundary C_k. There is a subsequence z_{k_n} of the sequence z_k which converges to a point z of the unit square. Now derive a contradiction as follows. Since f is differentiable at z,

$$f(w) = f(z) + f'(z)(w - z) + r(w)$$

where

$$|r(w)(w - z)^{-1}| \to 0 \quad \text{as} \quad w \to z.$$

Therefore for each $\varepsilon > 0$ there is a $\delta > 0$ such that if C_0 is the boundary of a square with sides of length h lying in the disc $|w - z| < \delta$, then

$$\left| \int_{C_0} f(z)\,dz \right| < \varepsilon h^2.$$

§3. The Cauchy integral formula

There are many approaches to the principal results of the theory of holomorphic functions. The most elegant approach is through Cauchy's Theorem and its chief consequence, the Cauchy integral formula. We begin with a special case, which itself is adequate for most purposes.

Theorem 3.1. *Suppose f is holomorphic in an open set Ω. Suppose C is a circle or rectangle contained in Ω and such that all points enclosed by C are in Ω. Then if w is enclosed by C,*

(1) $$f(w) = \frac{1}{2\pi i} \int_C f(z)(z - w)^{-1}\,dz.$$

(Here the integral is taken in the counterclockwise direction on C.)

Proof. Let $\gamma_0 : [a, b] \to \Omega$ be a piecewise smooth closed curve whose image is C, traced once in the counterclockwise direction. Given any positive ε which is so small that C encloses the closed disk of radius ε and center w, let C_ε be the circle of radius ε centered at w.

We can find a piecewise smooth curve $\gamma_1 \colon [a, b] \to \Omega$ which traces out C_ε once in the counterclockwise direction and is homotopic to γ_0 in the region Ω with the point w removed. Granting this for the moment, let us derive (1). By Exercise 1 of §2, and Cauchy's Theorem applied to $g(z) = f(z)(z - w)^{-1}$, we have

$$\int_C f(z)(z - w)^{-1}\, dz = \int_{\gamma_0} g = \int_{\gamma_1} g = \int_{C_\varepsilon} f(z)(z - w)^{-1}\, dz,$$

so

$$(2) \quad \int_C f(z)(z - w)^{-1}\, dz = \int_{C_\varepsilon} f(w)(z - w)^{-1}\, dz + \int_{C_\varepsilon} [f(w) - f(z)](z - w)^{-1}\, dz.$$

By Exercise 3 of §2, the first integral on the right in (2) equals $2\pi i f(w)$. Since f is differentiable, the integrand in the second integral on the right is bounded as $\varepsilon \to 0$. But the integration takes place over a curve of length $2\pi\varepsilon$, so this integral converges to zero as $\varepsilon \to 0$. Therefore (1) is true.

Finally, let us construct the curve γ_1 and the homotopy. For $t \in [a, b]$, let $\gamma_1(t)$ be the point at which C_ε and the line segment joining w to $\gamma_0(t)$ intersect. Then for $0 \le s \le 1$, let

$$\Gamma(t, s) = (1 - s)\gamma_0(t) + s\gamma_1(t).$$

It is easily checked that γ_0 and Γ have the desired properties. □

The preceding proof applies to any situation in which a given curve γ_1 is homotopic to all small circles around $w \in \Omega$. Let us make this more precise. A closed curve γ in an open set Ω is said to *enclose* the point $w \in \Omega$ *within* Ω if the following is true: there is a $\delta > 0$ such that if $0 < \varepsilon \le \delta$ then γ is homotopic, in Ω with w removed, to a piecewise smooth curve which traces out once, in a counterclockwise direction, the circle with radius ε and center w. We have the following generalization of Theorem 3.1.

Theorem 3.2. *Suppose f is holomorphic in an open set Ω. Suppose γ is a piecewise smooth closed curve in Ω which encloses a point w within Ω. Then*

$$(3) \qquad\qquad f(w) = \frac{1}{2\pi i} \int_\gamma f(z)(z - w)^{-1}\, dz.$$

Equation (3) is the *Cauchy integral formula*, and equation (1) is essentially a special case of (3). (In section 7 we shall give another proof of (1) when C is a circle.)

The Cauchy integral formula makes possible a second proof of the result of Exercise 5, §1: any holomorphic function can be represented locally as a power series.

Corollary 3.3. *Suppose f is holomorphic in Ω, and suppose Ω contains the disc*

$$\{z \mid |z - z_0| < R\}.$$

Then there is a unique power series such that

$$(4) \qquad f(w) = \sum_{n=0}^{\infty} a_n(w - z_0)^n, \qquad all \ w \ with \ |w - z_0| < R.$$

Proof. Suppose $0 < r < R$, and let C be the circle of radius r centered at z_0, with the counterclockwise direction. If $|w - z_0| < r$ and $z \in C$ then

$$|(w - z_0)(z - z_0)^{-1}| = s < 1.$$

We expand $(z - w)^{-1}$ in a power series:

$$z - w = z - z_0 - (w - z_0) = (z - z_0)[1 - (w - z_0)(z - z_0)^{-1}],$$

so

$$(z - w)^{-1} = (z - z_0)^{-1} \sum_{n=0}^{\infty} \left(\frac{w - z_0}{z - z_0} \right)^n.$$

The sequence of functions

$$g_N(z) = f(z) \sum_{n=0}^{N} (w - z_0)^n (z - z_0)^{-n-1}$$

converges uniformly for $z \in C$ to the function

$$f(z)(z - w)^{-1}.$$

Therefore we may substitute in equation (1) to get (4) with

$$(5) \qquad a_n = \frac{1}{2\pi i} \int_C f(z)(z - z_0)^{-n-1} \, dz.$$

This argument shows that the series exists and converges for $|w - z_0| < r$. The series is unique, since repeated differentiation shows that

$$(6) \qquad n! \, a_n = f^{(n)}(z_0).$$

Since $r < R$ was arbitrary, and since the series is unique, it follows that it converges for all $|w - z_0| < R$. ☐

Note that our two expressions for a_n can be combined to give

$$(7) \qquad f^{(n)}(z_0) = \frac{n!}{2\pi i} \int_C f(z)(z - z_0)^{-n-1} \, dz.$$

This is a special case of the following generalization of the Cauchy integral formula.

Corollary 3.4. *Suppose f is holomorphic in an open set containing a circle or rectangle C and all the points enclosed by C. If w is enclosed by C then the nth derivative of f at w is given by*

$$(8) \qquad f^{(n)}(w) = \frac{n!}{2\pi i} \int_C f(z)(z - w)^{-1} \, dz.$$

Proof. Let C_1 be a circle centered at w and enclosed by C. Then by the Cauchy integral theorem, and the argument given in the proof of Theorem 3.1, we may replace C by C_1 in (8). But in this case the formula reduces to the case given in (7). □

A function f defined and holomorphic on the whole plane $\mathbb{C}$ is said to be *entire*. The following result is known as *Liouville's Theorem*.

Corollary 3.5. *If f is an entire function which is bounded, then f is constant.*

Proof. We are assuming that there is a constant M such that $|f(z)| \leq M$, all $z \in \mathbb{C}$. It is sufficient to show that $f' \equiv 0$. Given $w \in \mathbb{C}$ and $R > 0$, let C be the circle with radius R centered at w. Then

$$f'(w) = \frac{1}{2\pi i} \int_C f(z)(z - w)^{-2} \, dz,$$

so

$$|f'(w)| \leq \frac{1}{2\pi} \cdot M \cdot R^{-2} \cdot 2\pi R = MR^{-1}.$$

Letting $R \to 0$ we get $f'(w) = 0$. □

A surprising consequence of Liouville's Theorem is the "Fundamental Theorem of Algebra."

Corollary 3.6. *Any nonconstant polynomial with complex coefficients has a complex root.*

Proof. Suppose p is such a polynomial. We may assume the leading coefficient is 1:

$$p(z) = z^n + a_{n-1}z^{n-1} + \cdots + a_1(z) + a_0.$$

It is easy to show that there is an $R > 0$ such that

(8) $|p(z)| \geq \frac{1}{2}|z|^n$ if $|z| \geq R$.

Now suppose p has no roots: $p(z) \neq 0$, all $z \in \mathbb{C}$. Then $f(z) = p(z)^{-1}$ would be an entire function. Then f would be bounded on the disc $|z| \leq R$, and (8) shows that it would be bounded by $2R^{-n}$ for $|z| \geq R$. But then f would be constant, a contradiction. □

Exercises

1. Verify the Cauchy Integral Formula in the form (1) by direct computation when $f(z) = e^z$ and C is a circle.

2. Compute the power series expansion (4) in the following cases. (Hint: (6) is not always the simplest way to obtain the a_n.)

   (a) $f(z) = \sin z$, $z_0 = \frac{1}{2}\pi$.

   (b) e^z, z_0 arbitrary.

(c) $f(z) = z^3 - 2z^2 + z + i$, $z_0 = -i$.

(d) $f(z) = (z - 1)(z^2 + 1)^{-1}$, $z_0 = 2$.

3. Derive equation (8) directly from (1) by differentiating.

4. Suppose f is an entire function, and suppose there are constants M and n such that

$$|f(z)| \leq M(1 + |z|)^n, \quad \text{all } z \in \mathbb{C}.$$

Show that f is a polynomial of degree $\leq n$. Show that, conversely, any polynomial of degree $\leq n$ satisfies such an inequality.

5. The Cauchy integral formula can be extended to more general situations, such as the case of a region bounded by more than one curve. For example, suppose Ω contains the annulus

$$A = \{z \mid r < |z - z_0| < R\},$$

and also the two circles

$$C_1 = \{z \mid |z - z_0| = r\}, \qquad C_2 = \{z \mid |z - z_0| = R\}$$

which bound A. Give C_1 and C_2 the counterclockwise direction. Then if $w \in A$, and f is holomorphic in Ω, show that

(*) $\qquad f(w) = \dfrac{1}{2\pi i} \displaystyle\int_{C_2} f(z)(z - w)^{-1} \, dz - \dfrac{1}{2\pi i} \int_{C_1} f(z)(z - w)^{-1} \, dz.$

(Hint: choose $a \in C$, $|a| = 1$ so that w does not lie on the line segment $L = \{z_0 + ta \mid r \leq t \leq R\}$ joining C_1 to C_2. There is a curve γ tracing out L, then C_2 in the counterclockwise direction, then L in the reverse direction, then C_1 in the clockwise direction. This curve is homotopic in Ω to any small circle about w. Moreover, the integral of $f(z)(z - w)^{-1}$ over γ equals the right side of (*), since the two integrations over L are in the opposite directions and cancel each other.)

6. Extend the result of Exercise 5 to the following situation: Ω contains a circle or rectangle C, together with all points enclosed by C *except* $z_1, z_2, \ldots, z_m$. Let $C_1, C_2, \ldots, C_m$ be circles around these points which do not intersect each other or C. If f is holomorphic in Ω and $w \in \Omega$ is enclosed by C, then

$$f(w) = \frac{1}{2\pi i} \int_C f(z)(z - w)^{-1} \, dz - \sum_{j=1}^{m} \frac{1}{2\pi i} \int_{C_j} f(z)(z - w)^{-1} \, dz;$$

again, all integrals are taken in the counterclockwise direction.

7. Suppose $f \colon \Omega \to \mathbb{C}$ is merely assumed to be differentiable at each point of the open set Ω, and suppose Ω contains a rectangle C and all points enclosed by C. Suppose w is enclosed by C. Modify the argument of Exercise 8, §2, to show that the integral in (1) remains unchanged if we replace C by any rectangle enclosed by C and enclosing w. Therefore show that (1) holds in this case also.

8. Use the result of Exercise 7 to show that if f is differentiable at each point of an open set Ω, the derivative is *necessarily* a continuous function.

§4. The local behavior of a holomorphic function

In this section we investigate the qualitative behavior near a point $z_0 \in \mathbb{C}$ of a function which is holomorphic in a disc around z_0. If f is not constant, then its qualitative behavior near z_0 is the same as that of a function of the form

$$f(z) = a_0 + a_m(z - z_0)^m$$

where a_0 and a_m are constants, $a_m \neq 0$, and $m \geq 1$.

Lemma 4.1. *Suppose f is holomorphic in an open set Ω and $z_0 \in \Omega$. If f is not constant near z_0, then*

(1) $$f(z) = a_0 + a_m(z - z_0)^m h(z)$$

where a_0 and a_m are constants, $m \geq 1$, and h is holomorphic in Ω with $h(z_0) = 1$.

Proof. Near z_0, f is given by a power series expansion

(2) $$f(z) = a_0 + a_1(z - z_0) + \cdots + a_n(z - z_0)^n + \cdots.$$

Let m be the first integer ≥ 1 such that $a_m \neq 0$. Then (2) gives (1) with

$$h(z) = \sum_{n = m}^{\infty} a_n a_m^{-1}(z - z_0)^{n - m}.$$

This function is holomorphic near z_0, and $h(z_0) = 1$. On the other hand, (1) defines a function h in Ω except at z_0, and the function so defined is holomorphic. Thus there is a single such function holomorphic throughout Ω. $\square$

Our first theorem here is the *Inverse Function Theorem* for holomorphic functions.

Theorem 4.2. *Suppose f is holomorphic in an open set Ω, and suppose $z_0 \in \Omega, f'(z_0) \neq 0$. Let $w_0 = f(z_0)$. Then there is an $\varepsilon_1 > 0$ and a holomorphic function g defined on the disc $|w - w_0| < \varepsilon_1$ such that*

$$g(\{w \mid |w - w_0| < \varepsilon_1\}) \text{ is open,}$$
$$f(g(w)) = w \qquad \text{if } |w - w_0| < \varepsilon_1.$$

In other words, f takes an open set containing z_0 in a 1-1 way onto a disc about w_0, and the inverse function g is holomorphic.

Proof. We begin by asking: Suppose the theorem were true. Can we derive a formula for g in terms of f? The idea is to use the Cauchy integral formula for g, using a curve γ around w_0 which is the image by f of a curve around z_0, because then we may take advantage of the fact that $g(f(z)) = z$. To carry this out, let $\delta > 0$ be small enough that Ω contains the closed disc $|z - z_0| \leq \delta$; later we shall further restrict δ. Let

$$\gamma_0(t) = z_0 + \delta e^{it}, \qquad t \in [0, 2\pi],$$
$$\gamma(t) = f(\gamma_0(t)),$$

and let C be the circle of radius δ around z_0. Assuming the truth of the theorem and assuming that γ enclosed w_1, we should have

$$g(w_1) = \frac{1}{2\pi i} \int_\gamma g(w)(w - w_1)^{-1}$$

$$= \frac{1}{2\pi i} \int_0^{2\pi} g(\gamma(t))\gamma'(t)[\gamma(t) - w_1]^{-1}\, dt$$

$$= \frac{1}{2\pi i} \int_0^{2\pi} \gamma_0(t)f'(\gamma_0(t))\gamma_0'(t)[f(\gamma_0(t)) - w_1]^{-1}\, dt$$

or

(3)
$$g(w_1) = \frac{1}{2\pi i} \int_C zf'(z)[f(z) - w_1]^{-1}\, dz.$$

Our aim now is to use (3) to *define* g and show that it has the desired properties. First, note that (1) holds with $m = 1$. We may restrict δ still further, so that $|h(z)| \geq \frac{1}{2}$ for $z \in C$. This implies that $f(z) \neq w_0$ if $z \in C$. Then we may choose $\varepsilon > 0$ so that

$$f(z) \neq w_1 \qquad \text{if } |w_1 - w_0| < \varepsilon \text{ and } z \in C.$$

With this choice of δ and ε, (3) defines a function g on the disc $|w_1 - w_0| < \varepsilon$. This function is holomorphic; in fact it may be differentiated under the integral sign.

Suppose

$$|z_1 - z_0| < \delta, \qquad f(z_1) = w_1, \quad \text{and} \quad |w_1 - w_0| < \varepsilon.$$

We can, and shall, assume that δ is chosen so small that $f'(z) \neq 0$ when $|z - z_0| < \delta$. Then

$$f(z) - w_1 = f(z) - f(z_1) = (z - z_1)k(z)$$

where k is holomorphic in Ω and $k(z_1) = f'(z_1) \neq 0$. Therefore k is nonzero in Ω. We have

$$g(w_1) = \frac{1}{2\pi i} \int_C zf'(z)k(z)^{-1}(z - z_1)^{-1}\, dz.$$

But the right side is the Cauchy integral formula for

$$z_1 f'(z_1)k(z_1)^{-1} = z_1.$$

Thus $g(f(z_1)) = z_1$ for z_1 near z_0, and we have shown that f is 1-1 near z_0. Also

$$1 = (g \circ f)'(z_0) = g'(w_0)f'(z_0),$$

so $g'(w_0) \neq 0$. Therefore g is 1-1 near w_0. We may take $\varepsilon_0 > 0$ so small that $\varepsilon_0 \leq \varepsilon$, and so that g is 1-1 on the disc $|w - w_0| < \varepsilon_0$. We may also assume $\varepsilon_1 \leq \varepsilon_0$ so small that

$$|w - w_0| < \varepsilon_1 \quad \text{implies} \quad |g(w) - z_0| < \delta$$

and
$$|f(g(w)) - w_0| < \varepsilon_0.$$
But then for $|w - w_0| < \varepsilon_1$ we have
$$g(f(g(w))) = g(w),$$
and since g is 1-1 on $|w - w_0| < \varepsilon_0$ this implies
$$f(g(w)) = w. \qquad \square$$

As an example, consider the logarithm. Suppose $w_0 \in \mathbb{C}$, $w_0 \neq 0$. We know by results of §6 of Chapter 2 that there is a $z_0 \in \mathbb{C}$ such that $e^{z_0} = w_0$. The derivative of e^z at z_0 is $e^{z_0} = w_0 \neq 0$. Therefore there is a unique way of choosing the logarithm
$$z = \log w,$$
z near z_0, in such a way that it is a holomorphic function of w near w_0. (In fact, we know that any two determinations of $\log w$ differ by an integral multiple of $2\pi i$; therefore the choice of $\log w$ will be holomorphic in an open set Ω if and only if it is continuous there.) By definition a *branch* of the logarithm function in Ω is a choice $z = \log w$, $w \in \Omega$, such that z is a holomorphic function of w in Ω.

As a second example, consider the nth root, n a positive integer. If $w_0 \neq 0$, choose a branch of $\log w$ holomorphic in a disc about w_0. Then if we set
$$w^{1/n} = \exp\left(\frac{1}{n}\log w\right),$$
this is a holomorphic function of w near w_0 and
$$(w^{1/n})^n = \exp\left(n \cdot \frac{1}{n}\log w\right) = \exp(\log w) = w.$$

We refer to $w^{1/n}$ as a branch of the nth root.

There are exactly n branches of the nth root holomorphic near w_0. In fact, suppose
$$z_0{}^n = w_0 = z_1{}^n.$$
Then for any choice of $\log z_0$ and $\log z_1$,
$$\exp(n \log z_0) = \exp(n \log z_1),$$
so
$$n \log z_0 = n \log z_1 + 2m\pi i,$$
some integer m. This implies

(4) $$z_0 = \exp(\log z_0) = \cdots = z_1 \exp(2m\pi i n^{-1}).$$

Since
$$\exp(2m\pi i n^{-1}) = \exp(2m'\pi i n^{-1})$$
if and only if $(m - m')n^{-1}$ is an integer, we get all n distinct nth roots of w_0 by letting z_1 be a fixed root and taking $m = 0, 1, \ldots, n - 1$ in (4).

We can now describe the behavior of a nonconstant holomorphic function near a point where the derivative vanishes.

Theorem 4.3. *Suppose f is holomorphic in an open set Ω, and suppose $z_0 \in \Omega$. Suppose f is not constant near z_0, and let m be the first integer ≥ 1 such that $f^{(m)}(z_0) \neq 0$. Let $w_0 = f(z_0)$. There are $\varepsilon > 0$ and $\delta > 0$ such that if*

$$0 < |w - w_0| < \varepsilon$$

there are exactly m distinct points z such that

$$|z - z_0| < \delta \quad \text{and} \quad f(z) = w.$$

Proof. By Lemma 4.1,

$$f(z) = w_0 + (z - z_0)^m h(z),$$

where h is holomorphic in Ω and $h(z_0) \neq 0$. Choose a branch g of the mth root function which is holomorphic near $h(z_0)$. Then near z_0,

$$f(z) = w_0 + [(z - z_0)(g(h(z)))]^m = w_0 + k(z)^m,$$

where k is holomorphic near z_0. Then $k(z_0) = 0$, $k'(z_0) = g(h(z_0)) \neq 0$. For z near z_0, $z \neq z_0$, we have

$$f(z) = w \quad \text{if and only if} \quad k(z) = (w - w_0)^{1/m}$$

for some determination of the mth root of $w - w_0$. We can apply Theorem 4.2 to k: there are ε, δ so that

$$k(z) = t$$

has a unique solution z in the disc $|z - z_0| < \delta$ for each t in the disc $|t| < \varepsilon^{1/m}$. But if

$$0 < |w - w_0| < \varepsilon$$

then $w - w_0$ has exactly m mth roots t, all with $|t| < \varepsilon^{1/m}$. $\square$

The following corollary is called the *open mapping property* of holomorphic functions.

Corollary 4.4. *If Ω is open and $f: \Omega \to \mathbb{C}$ is holomorphic and not constant near any point, then $f(\Omega)$ is open.*

Proof. Suppose $w_0 \in f(\Omega)$. Then $w_0 = f(z_0)$, some $z_0 \in \Omega$. We want to show that there is an $\varepsilon > 0$ such that the disc $|w - w_0| < \varepsilon$ is contained in $f(\Omega)$. But this follows from Theorem 4.3. $\square$

Exercises

1. Use Corollary 4.4 to prove the *Maximum Modulus Theorem*: if f is holomorphic and not constant in a disc $|z - z_0| < R$, then $g(z) = |f(z)|$ does not have a local maximum at z_0.

2. With f as in Exercise 1, show that $g(z) = |f(z)|$ does not have a local minimum at z_0 unless $f(z_0) = 0$.

3. Use Exercise 2 to give another proof of the Fundamental Theorem of Algebra.

4. Suppose $z = \log w$. Show that Re $z = \log |w|$.

5. Suppose f is holomorphic near z_0 and $f(z_0) \neq 0$. Show that $\log |f(z)|$ is harmonic near z_0.

6. Use Exercise 5 and the maximum principle for harmonic functions to give another proof of the Maximum Modulus Theorem.

7. Use the Cauchy integral formula (for a circle with center z_0) to give still another proof of the Maximum Modulus Theorem.

8. Use the Maximum Modulus Theorem to prove Corollary 4.4. (Hint: let $w_0 = f(z_0)$ and let C be a small circle around z_0 such that $f(z) \neq w_0$ if $z \in C$. Choose $\varepsilon > 0$ so that $|f(z) - w_0| \geq 2\varepsilon$ if $z \in C$. If $|w - w_0| < \varepsilon$, can $(f(z) - w)^{-1}$ be holomorphic inside C?)

9. A set $\Omega \subset \mathbb{C}$ is *connected* if for any points $z_0, z_1 \in \Omega$ there is a (continuous) curve $\gamma: [a, b] \to \Omega$ with $\gamma(a) = z_0$, $\gamma(b) = z_1$. Suppose Ω is open and connected, and suppose f is holomorphic in Ω. Show that if f is identically zero in any nonempty open subset $\Omega_1 \subset \Omega$, then $f \equiv 0$ in Ω.

10. Let Ω be the union of two disjoint open discs. Show that Ω is not connected.

§5. Isolated singularities

Suppose f is a function holomorphic in an open set Ω. A point z_0 is said to be an *isolated singularity* of f if $z_0 \notin \Omega$ but if every point sufficiently close to z_0 is in Ω. Precisely, there is a $\delta > 0$ such that

$$z \in \Omega \qquad \text{if } 0 < |z - z_0| < \delta.$$

For example, 0 is an isolated singularity for $f(z) = z^{-n}$, n a positive integer, and for $g(z) = \exp(1/z)$. On the other hand, according to the definition, 0 is also an isolated singularity for the function f which is defined by $f(z) = 1$, $z \neq 0$ and is not defined at 0. This example shows that a singularity may occur through oversight: not assigning values to enough points. An isolated singularity z_0 for f is said to be a *removable singularity* if f can be defined at z_0 in such a way as to remain holomorphic.

Theorem 5.1. *Suppose z_0 is an isolated singularity for the holomorphic function f. It is a removable singularity if and only if f is bounded near z_0, i.e., there are constants M, $\delta > 0$ such that*

$$(1) \qquad\qquad |f(z)| \leq M \qquad \text{if } 0 < |z - z_0| < \delta.$$

Proof. Suppose z_0 is a removable singularity. Then f has a limit at z_0, and it follows easily that (1) is true.

Conversely, suppose (1) is true. Choose r with $0 < r < \delta$ and let $\varepsilon > 0$ be such that $0 < \varepsilon < r$. Let C be the circle with center z_0 and radius r. Given w with $0 < |w - z_0| < r$, choose ε so small that $0 < \varepsilon < |w - z_0|$, and let C_ε be the circle with center z_0 and radius ε. By Exercise 5 of §3,

$$f(w) = \frac{1}{2\pi i} \int_C f(z)(z - w)^{-1}\, dz - \frac{1}{2\pi i} \int_{C_\varepsilon} f(z)(z - w)^{-1}\, dz.$$

Since f is bounded on C independent of ε, the second integral goes to zero as $\varepsilon \to 0$. Thus

(2) $$f(w) = \frac{1}{2\pi i} \int_C f(z)(z - w)^{-1}\, dz, \qquad 0 < |w - z_0| < r.$$

We may *define* $f(z_0)$ by (2) with $w = z_0$, and then (2) will hold for all w, $|w - z_0| < r$. The resulting function is then holomorphic. □

An isolated singularity z_0 for a function f is said to be a *pole of order n* for f, where n is an integer ≥ 1, if f is of the form

(3) $$f(z) = (z - z_0)^{-n} g(z)$$

where g is defined at z_0 and holomorphic near z_0, while $g(z_0) \neq 0$. A pole of order 1 is often called a *simple pole*.

Theorem 5.2. *Suppose z_0 is an isolated singularity for the holomorphic function f. It is a pole of order n if and only if the function*

$$(z - z_0)^n f(z)$$

is bounded near z_0, while the function

$$(z - z_0)^{n-1} f(z)$$

is not.

Proof. It follows easily from the definition that if z_0 is a pole of order n the asserted consequences are true.

Conversely, suppose $(z - z_0)^n f(z) = g(z)$ is bounded near z_0. Then z_0 is an isolated singularity, so we may extend g to be defined at z_0 and holomorphic. We want to show that $g(z_0) \neq 0$ if $(z - z_0)^{n-1} f(z)$ is not bounded near z_0. But if $g(z_0) = 0$ then by Lemma 4.1,

$$g(z) = (z - z_0)^m h(z)$$

for some $m \geq 1$ and some h holomorphic near z_0. But then $(z - z_0)^{n-1} f(z) = (z - z_0)^{m-1} h(z)$ is bounded near z_0. □

An isolated singularity which is neither removable nor a pole (of any order) is called an *essential singularity*. Note that if z_0 is a pole or a removable singularity, then for some $a \in \mathbb{C}$ *or* $a = \infty$

$$f(z) \to a \quad \text{as} \quad z \to z_0$$

This is most emphatically not true near an essential singularity.

Theorem 5.3 (Casorati-Weierstrass). *Suppose z_0 is an isolated singularity for the holomorphic function f. If z_0 is an essential singularity for f, then for any $\varepsilon > 0$ and any $a \in \mathbb{C}$ there is a z such that*

$$|z - z_0| < \varepsilon, \qquad |f(z) - a| < \varepsilon.$$

Proof. Suppose the conclusion is not true. Then for some $\varepsilon > 0$ and some $a \in \mathbb{C}$ we have

$$|f(z) - a| \geq \varepsilon \quad \text{where} \quad 0 < |z - z_0| < \varepsilon.$$

Therefore $h(z) = (f(z) - a)^{-1}$ is bounded near z_0. It follows that h can be extended so as to be defined at z_0 and holomorphic near z_0. Then for some $m \geq 0$,

$$h(z) = (z - z_0)^m k(z)$$

where k is holomorphic near z_0 and $k(z_0) \neq 0$. We have

$$f(z) = a + h(z)^{-1} = a + (z - z_0)^{-m} k(z)^{-1}, \qquad 0 < |z - z_0| < \varepsilon.$$

Therefore z_0 is either a removable singularity or a pole for f. $\quad\square$

Actually, much more is true. Picard proved that if z_0 is an isolated essential singularity for f, then for any $\varepsilon > 0$ and any $a \in \mathbb{C}$, with at most one exception, there is a z such that $0 < |z - z_0| < \varepsilon$ and $f(z) = a$. An example is $f(z) = \exp(1/z)$, $z \neq 0$, which takes any value except zero in any disc around zero.

Isolated singularities occur naturally in operations with holomorphic functions. Suppose, for example, that f is holomorphic in Ω and $z_0 \in \Omega$. If $f(z_0) \neq 0$, then we know that $f(z)^{-1}$ is holomorphic near z_0. The function f is said to have a *zero of order n* (or *multiplicity n*) at z_0, n an integer ≥ 0, if

$$\begin{aligned} f^{(k)}(z_0) &= 0, \qquad 0 \leq k < n. \\ f^{(n)}(z_0) &\neq 0. \end{aligned}$$

(In particular, f has a zero of order *zero* at z_0 if $f(z_0) \neq 0$.) A zero of order one is called a *simple zero*.

Lemma 5.4. *If f is holomorphic near z_0 and has a zero of order n at z_0, then $f(z)^{-1}$ has a pole of order n at z_0.*

Proof. By Lemma 4.1,

(4) $$f(z) = (z - z_0)^n h(z),$$

where h is holomorphic near z_0 and $h(z_0) \neq 0$. The desired conclusion follows. $\quad\square$

For example, the function

$$\sec z = (\cos z)^{-1}$$

is holomorphic except at the zeros of $\cos z$, where it has poles (of order 1). The same is then true of

$$\tan z = \sin z (\cos z)^{-1} = \sin z \sec z.$$

It is convenient to assign the "value" ∞ at z_0 to a holomorphic function with a pole at z_0. Similarly, if z_0 is a removable singularity for f we shall consider f as being extended to take the appropriate value at z_0. With these conventions we may work with the following extension of the notion of holomorphic function.

Suppose $\Omega \subset \mathbb{C}$ is an open set. A function $f \colon \Omega \to \mathbb{C} \cup \{\infty\}$ is said to be *meromorphic in* Ω if for each point $z_0 \in \Omega$, either z_0 is a pole of f, or f is holomorphic in the disc $|z - z_0| < \delta$ for some $\delta > 0$.

Theorem 5.5. *Suppose Ω is open and f, g are meromorphic in Ω. Suppose $a \in \mathbb{C}$. Then the functions*

$$af, \qquad f + g, \qquad fg$$

are meromorphic in Ω. If Ω is connected and $g \not\equiv 0$ in Ω, then f/g is meromorphic in Ω.

Proof. We leave all but the last statement as an exercise. Suppose Ω is connected, i.e. for each $z_0, z_1 \in \Omega$ there is a continuous curve $\gamma \colon [0, 1] \to \Omega$ with $\gamma(0) = z_0$, $\gamma(1) = z_1$. Given any $z_0 \in \Omega$, either $g(z)^{-1}$ is meromorphic near z_0 or g vanishes in a disc around z_0. We want to show that the second alternative implies $g \equiv 0$ in all of Ω.

Suppose g vanishes identically near $z_0 \in \Omega$ and suppose z_1 is any other point of Ω. Let γ be a curve joining z_0 to $z_1 \colon \gamma(0) = z_0$, $\gamma(1) = z_1$. Let A be the subset of the interval $[0, 1]$ consisting of all those t such that g vanishes identically near $\gamma(t)$. Let $c = \text{lub } A$. There is a sequence $(t_n)_1^\infty \subset A$ such that $t_n \to c$. If g did not vanish identically near $\gamma(c)$ then either $g(\gamma(c)) \neq 0$, or $\gamma(c)$ is a zero of order n for some n, or $\gamma(c)$ is a pole of order n for some n. But then (see (3) and (4)) we could not have g identically zero near $\gamma(t_n)$ for those n so large that $\gamma(t_n)$ is very close to $\gamma(c)$. Thus g vanishes identically near c. This means that $c = 1$, since otherwise g vanishes identically near $\gamma(c + \varepsilon)$ for small $\varepsilon > 0$. Therefore $g(z_1) = g(\gamma(1)) = g(\gamma(c)) = 0$.

We know now that given $z_0 \in \Omega$, either $g(z_0) \neq 0$ or z_0 is either a zero or pole of order n for g. It follows that either $g(z)^{-1}$ is holomorphic near z_0, or, by (4), that z_0 is a pole of order n, or, by (3), that z_0 is a zero of order n. Thus $g(z)^{-1}$ is meromorphic in Ω, and so f/g is also. $\square$

Exercises

1. Prove the rest of the assertions of Theorem 5.5.

2. Show that $\tan z$ is meromorphic on all of $\mathbb{C}$, with only simple poles. What about $(\tan z)^2$?

3. Determine all the functions f, meromorphic in all of $\mathbb{C}$, such that

$$|f(z) - \tan z| < 2$$

at each point z which is not a pole either of f or of $\tan z$.

4. Suppose f has a pole of order n at z_0. Show that there are an $R > 0$ and a $\delta > 0$ such that if $|w| > R$ then there are exactly n points z such that

$$f(z) = w, \qquad |z - z_0| < \delta.$$

(Hint: recall Theorem 4.3.)

5. A function f is said to be defined *near* ∞ if there is an $R > 0$ such that

$$\{z \mid |z| > R\}$$

is in the domain of definition of f. The function f is said to be *holomorphic at* ∞ if 0 is a removable singularity for the function g defined by

$$g(z) = f(1/z),$$

$1/z$ in the domain of f. Similarly, ∞ is said to be a zero or pole of order n for f if 0 is a zero or pole of order n for g. Discuss the status of ∞ for the following functions:

(a) $f(z) = z^n$, n an integer.
(b) $f(z) = e^z$.
(c) $f(z) = \sin z$.
(d) $f(z) = \tan z$.

6. Suppose z_0 is an essential singularity for f, while g is meromorphic near z_0 and not identically zero near z_0. Is z_0 an essential singularity for fg? What if, in addition, z_0 is not an essential singularity of g?

§6. Rational functions; Laurent expansions; residues

A *rational function* is the quotient of two polynomials:

$$f(z) = p(z)/q(z)$$

where p and q are polynomials, $q \neq 0$. By Theorem 5.4, a rational function is meromorphic in the whole plane $\mathbb{C}$. (In fact it is also meromorphic at ∞; see Exercise 5 of §5.)

It is easy to see that sums, scalar multiples, and products of rational functions are rational functions. If f and g are rational functions and $g \neq 0$, then f/g is a rational function. In particular, any function of the form

(1) $f(z) = a_1(z - z_0)^{-1} + a_2(z - z_0)^{-2} + \cdots + a_n(z - z_0)^{-n}$

is a rational function with a pole at z_0. We can write

(2) $f(z) = p((z - z_0)^{-1})$

where p is a polynomial with $p(0) = 0$. It turns out that any rational function is the sum of a polynomial and rational functions of the form (1).

Theorem 6.1. *Suppose f is a rational function with poles at the distinct points $z_1, z_2, \ldots, z_m$ and no other poles. There are unique polynomials p_0, $p_1, \ldots, p_m$ such that $p_j(0) = 0$ if $j \neq 0$ and*

(3) $f(z) = p_0(z) + p_1((z - z_1)^{-1}) + \cdots + p_m((z - z_m)^{-1}).$

Proof. We induce on m. If $m = 0$, then f has no poles. Thus f is an entire function. We have $f = p/q$ where p and q are polynomials and $q \not\equiv 0$. Suppose p is of degree r and q is of degree s. It is not hard to show that there is a constant a such that

$$z^{s-r}f(z) \to a \quad \text{as} \quad z \to \infty.$$

This and Exercise 4 of §3 imply that f is a polynomial.

Now suppose the assertion of the theorem is true for rational functions with $m - 1$ distinct poles, and suppose f has m poles. Let z_0 be a pole of f, of order r. Then

$$f(z) = (z - z_0)^{-r}h(z),$$

where h is holomorphic near z_0. Near z_0,

$$h(z) = \sum_{n=1}^{\infty} a_n(z - z_0)^n.$$

Therefore

$$f(z) = a_0(z - z_0)^{-r} + a_1(z - z_0)^{1-r} + \cdots + a_{r-1}(z - z_0)^{-1} + k(z),$$

where k is holomorphic near z_0. Now k is the difference of two rational functions, hence is rational. The function $g = f - k$ has no poles except at z_0, so the poles of k are the poles of f which differ from z_0. By the induction assumption, k has a unique expression of the desired form. Therefore f has *an* expression of the desired form.

Finally, we want to show that the expression (3) is unique. Suppose p_0 is of degree k. The coefficient b_k of z^k in p_0 can be computed by taking limits on both sides in (3):

$$\lim_{z \to \infty} z^{-k}f(z) = \lim_{z \to \infty} z^{-k}p_0(z) = b_k.$$

Therefore the coefficient of z^{k-1} can be computed:

$$\lim_{z \to \infty} z^{1-k}[f(z) - b_k z^k] = b_{k-1}, \quad \text{etc.}$$

Continuing in this way, we determine all coefficients of p_0. Similarly, if p_1 is of order r then the coefficient of the highest power of $(z - z_1)^{-1}$ in (3) is

$$\lim_{z \to z_1} (z - z_1)^r f(z) = c_r,$$

the coefficient of the next power is

$$\lim_{z \to z_1} (z - z_1)^{r-1}[f(z) - c_r(z - z_1)^{-r}].$$

All coefficients may be computed successively in this way.   □

The expression (3) is called the *partial fractions decomposition* of the rational function f.

Let us note explicitly a point implicit in the preceding proof. If f has a pole of order r at z_0, then

$$f(z) = \sum_{n=-r}^{\infty} b_n(z - z_0)^n, \qquad 0 < |z - z_0| < \delta,$$

some $\delta > 0$. This generalization of the power series expansion of a holomorphic function is called a *Laurent expansion*. It is one case of a general result valid, in particular, near any isolated singularity.

Theorem 6.2. *Suppose f is holomorphic in the annulus*

$$A = \{z \mid r < |z - z_0| < R\}.$$

Then there is a unique two sided sequence $(a_n)_{-\infty}^{\infty} \subset \mathbb{C}$ such that

(4) $$f(z) = \sum_{-\infty}^{\infty} a_n(z - z_0)^n, \qquad r < |z - z_0| < R.$$

Proof. Suppose $r < |z - z_0| < R$. Choose r_1, R_1 such that

$$r < r_1 < |z - z_0| < R_1 < R.$$

Let C_1 be the circle of radius r_1 and C_2 the circle of radius R_1 centered at z_0. By Exercise 5 of §3,

$$f(z) = \frac{1}{2\pi i} \int_{C_2} f(w)(w - z)^{-1}\, dw - \frac{1}{2\pi i} \int_{C_1} f(w)(w - z)^{-1}\, dw$$

$$= f_2(z) + f_1(z).$$

Here f_2 and f_1 are defined by the respective integrals. We consider f_2 as being defined for $|z - z_0| < R_1$ and f_1 as being defined for $|z - z_0| > r_1$. Then f_2 is holomorphic and has the power series expansion

(5) $$f_2(z) = \sum_{n=0}^{\infty} a_n(z - z_0)^n, \qquad |z - z_0| < R_1.$$

Moreover, by the Cauchy integral theorem we may increase R_1 without changing the values of f_2 on $|z - z_0| < R_1$; thus the series (5) converges for $|z - z_0| < R$.

The function f_1 is holomorphic for $|z - z_0| > r_1$. Again, by moving the circle C_1, we may extend f_1 to be holomorphic for $|z - z_0| > r$. To get an appropriate series expansion we proceed as in the proof of Corollary 3.3. When $|z - z_0| > r_1$ and $|w - z_0| = r_1$,

$$(w - z)^{-1} = ((w - z_0) - (z - z_0))^{-1} = -(z - z_0)^{-1}[1 - (w - z_0)(z - z_0)^{-1}]^{-1}$$

$$= -\sum_{n=1}^{\infty} (w - z_0)^{n-1}(z - z_0)^{-n}.$$

The series converges uniformly for $w \in C_1$, so

(6) $$f_1(z) = \sum_{n=1}^{\infty} a_{-n}(z - z_0)^{-n}$$

where

$$a_{-n} = \frac{1}{2\pi i} \int_{C_1} f(w)(w - z_0)^{n-1} \, dw.$$

Equations (5) and (6) together give (4).

Finally, we want to prove uniqueness. Suppose (4) is valid. Then the power series

$$\sum_{n=-\infty}^{\infty} a_n(z - z_0)^n$$

converges for $r < |z - z_0| < R$, so it converges *uniformly* on any smaller annulus. It follows that if C is any circle with center z_0, contained in A, then (3) may be integrated term by term over C. Since

$$\int_C (z - z_0)^n \, dz$$

is zero for $n \neq 1$ and $2\pi i$ for $n = -1$, this gives

(7)
$$a_{-1} = \frac{1}{2\pi i} \int_C f(z) \, dz.$$

More generally we may multiply f by $(z - z_0)^{-m-1}$ and integrate to get

(8)
$$a_m = \frac{1}{2\pi i} \int_C f(z)(z - z_0)^{-m-1} \, dz,$$

all m. Thus the coefficients are uniquely determined. □

In particular, Theorem 6.2 applies when f has an isolated singularity at z_0. In this case the Laurent expansion (3) is valid for

$$0 < |z - z_0| < R,$$

some $R > 0$. The coefficient a_{-1} is called the *residue* of f at z_0. Equation (7) determines the residue by evaluating an integral; reversing the viewpoint we may evaluate the integral if we can determine the residue. These observations are the basis for the "calculus of residues." The following theorem is sufficient for many applications.

Theorem 6.3. *Suppose C is a circle or rectangle. Suppose f is holomorphic in an open set Ω containing C and all points enclosed by C, except for isolated singularities at the points $z_1, z_2, \ldots, z_m$ enclosed by C. Suppose the residue of f at z_j is b_j. Then*

(9)
$$\int_C f(z) \, dz = 2\pi i(b_1 + b_2 + \cdots + b_m).$$

Proof. Let $C_1, \ldots, C_m$ be nonoverlapping circles centered at $z_1, \ldots, z_m$ and enclosed by C. Then

$$2\pi i b_j = \int_{C_j} f(z) \, dz.$$

Applying Exercise 6 of §3, we get (9).   □

If f has a pole at z_0, the residue may be computed as follows. If n is the order of the pole,

$$f(z) = (z - z_0)^{-n} h(z) = (z - z_0)^{-n} \sum_{m=0}^{\infty} b_m (z - z_0)^m,$$

so the residue at z_0 is

$$b_{n-1} = [(n - 1)!]^{-1} h^{(n-1)}(z_0).$$

In particular, at a simple pole $n = 1$ and the residue is $h(z_0)$.

Let us illustrate the use of the calculus of residues by an example. Suppose we want to compute

$$I = \int_0^{\infty} (1 + t^2)^{-1}\, dt$$

and have forgotten that $(1 + t^2)^{-1}$ is the derivative of $\tan^{-1} t$. Now

$$I = \frac{1}{2} \int_{-\infty}^{\infty} (1 + t^2)^{-1}\, dt.$$

Let C_R, $R > 0$, be the square with vertices $\pm R$ and $\pm R + Ri$. Let $f(z) = (1 + z^2)^{-1}$. The integral of f over the three sides of C_R which do not lie on the real axis is easily seen to approach zero as $R \to \infty$. Therefore

$$\int_{-\infty}^{\infty} (1 + t^2)^{-1}\, dt = \lim_{r \to \infty} \int_{-R}^{R} (1 + t^2)^{-1}\, dt$$

$$= \lim_{R \to \infty} \int_{C_R} f(z)\, dz.$$

For $R > 1$, f is holomorphic inside C_R except at $z = i$, where it has a simple pole. Since

$$f(z) = (z - i)^{-1}(z + i)^{-1},$$

the residue at i is $(2i)^{-1}$. Therefore when $R > 1$,

$$\int_{C_R} f(z)\, dz = 2\pi i (2i)^{-1} = \pi.$$

We get $I = \frac{1}{2}\pi$ (which is $\tan^{-1}(+\infty) - \tan^{-1} 0$, as it should be).

We conclude with some further remarks on evaluating integrals by this method. Theorem 6.3 is easily shown to be valid for other curves C, such as a semicircular arc together with the line segment joining its endpoints, or a rectangle with a portion or portions replaced by semicircular arcs. The method is of great utility, depending on the experience and ingenuity of the user.

Exercises

1. Compute the partial fractions decomposition of

$$z^3(z^2 - 3z + 2)^{-1}, \qquad (z^3 + 2z^2 + z)^{-1}.$$

2. Find the Laurent expansion of $\exp(1/z)$ and $\sin(1/z)$ at 0.
3. Compute the definite integrals

$$\int_{-\infty}^{\infty} (t + 1)(t^3 - 2it^2 + t - 2i)^{-1} dt$$

$$\int_{0}^{\infty} t^2(t^4 + 1)^{-1} dt.$$

4. Show that $\int_0^\infty t^{-1} \sin t \, dt = \frac{1}{2}\pi$. (Hint: this is an even function, and it is the imaginary part of $f(z) = z^{-1}e^{iz}$. Integrate f over rectangles lying in the half-plane $\operatorname{Im} z \geqslant 0$, but with the segment $-\varepsilon < t < \varepsilon$ replaced by a semicircle of radius ε in the same half-plane, and let the rectangles grow long in proportion to their height.)
5. Show that a rational function is holomorphic at ∞ or has a pole at ∞.
6. Show that any function which is meromorphic in the whole plane and is holomorphic at ∞, or has a pole there, is a rational function.
7. Show that if $\operatorname{Re} z > 0$, the integral

$$\Gamma(z) = \int_{0}^{\infty} t^{z-1}e^{-t} dt$$

exists and is a holomorphic function of z for $\operatorname{Re} z > 0$. This is called the *Gamma function*.
8. Integrate by parts to show that

$$\Gamma(z + 1) = z\Gamma(z), \qquad \operatorname{Re} z > 0.$$

9. Define $\Gamma(z)$ for $-1 < \operatorname{Re} z \leq 0, z \neq 0$, by

$$\Gamma(z) = z^{-1}\Gamma(z + 1).$$

Show that Γ is meromorphic for $\operatorname{Re} z > -1$, with a simple pole at zero.
10. Use the procedure of Exercise 9 to extend Γ so as to be meromorphic in the whole plane, with simple poles at $0, -1, -2, \ldots$.
11. Show inductively that the residue of Γ at $-n$ is $(-1)^n(n!)^{-1}$.
12. Is Γ a rational function?

§7. Holomorphic functions in the unit disc

In this section we discuss functions holomorphic in the unit disc $D = \{z \mid |z| < 1\}$ from the point of view of periodic functions and distributions. This point of view gives another way of deriving the basic facts about the local theory of holomorphic functions. It also serves to introduce certain *spaces* of holomorphic functions and of periodic distributions.

Note that if f is defined and holomorphic for $|z - z_0| < R$, then setting

$$f_1(w) = f(z_0 + Rw)$$

we get a function f_1 holomorphic for $|w| < 1$. Similarly, if f has an isolated singularity at z_0, we may transform it to a function with an isolated singularity at zero and holomorphic elsewhere in the unit disc. Since f can be recovered from f_1 by

$$f(z) = f_1(R^{-1}(z - z_0)),$$

all the information about local behavior can be deduced from study of f_1 instead.

Suppose f is holomorphic in D. Then the function

$$g(r, \theta) = f(re^{i\theta})$$

is periodic as a function of θ for $0 \leq r < 1$ and constant for $r = 0$. It is also differentiable, and the assumption that f is holomorphic imposes a condition on the derivatives of g. In fact

$$h^{-1}[g(r + h, \theta) - g(r, \theta)] = h^{-1}[f(re^{i\theta} + he^{i\theta}) - f(re^{i\theta})]$$
$$= e^{i\theta}(he^{i\theta})^{-1}[f(re^{i\theta} + he^{i\theta}) - f(re^{i\theta})].$$

Letting $h \to 0$ we get

$$\frac{\partial}{\partial r} g(r, \theta) = e^{i\theta} f'(re^{i\theta}).$$

Similarly,

$$h^{-1}[g(r, \theta + h) - g(r, \theta)] = h^{-1}[f(re^{i(\theta + h)}) - f(re^{i\theta})]$$
$$\sim f'(re^{i\theta}) \cdot h^{-1}[re^{i(\theta + h)} - re^{i\theta}]$$

so

$$\frac{\partial}{\partial \theta} g(r, \theta) = ire^{i\theta} f'(re^{i\theta}).$$

Combining these equations we get

(1) $$ir \frac{\partial g}{\partial r} = \frac{\partial g}{\partial \theta}.$$

Now let $g_r(\theta) = g(r, \theta)$, $0 \leq r < 1$. Since g_r is continuous, periodic, and continuously differentiable as a function of θ, it is the sum of its Fourier series:

(2) $$g_r(\theta) = \sum_{-\infty}^{\infty} a_n(r)e^{in\theta}.$$

The coefficients $a_n(r)$ are given by

$$a_n(r) = \frac{1}{2\pi} \int_0^{2\pi} g(r, \theta)e^{-in\theta} \, d\theta.$$

It follows that $a_n(r)$ is continuous for $0 \leq r < 1$ and differentiable for $0 < r < 1$, with derivative

$$a_n'(r) = \frac{1}{2\pi} \int_0^{2\pi} \frac{\partial g}{\partial r}(r, \theta) e^{-in\theta} \, d\theta.$$

Using (1) and integrating by parts we get

(3) $a_n'(r) = r^{-1} n a_n(r), \qquad n \neq 0,$

and $a_0'(r) = 0$. Thus a_0 is constant. The equation (3) may be solved for a_n as follows, $n \neq 0$. The real and imaginary parts of a_n are each real solutions of

$$u'(r) = r^{-1} n u(r).$$

On any interval where $u(r) \neq 0$ this is equivalent to

$$\frac{d}{dr} \log |u(r)| = \frac{d}{dr} \log r^n$$

so on such an interval $u(r) = cr^n$, c constant. Since a_n is continuous on $[0, 1)$ and vanishes at 0 if $n \neq 0$ (because g_0 is constant), we must have $a_n(r) = a_n r^n$, with a_n constant and $a_n = 0$ if $n < 0$.

We have proved the following: if f is holomorphic in D, then

(4) $f(re^{i\theta}) = \sum_{n=0}^{\infty} a_n r^n e^{in\theta}, \qquad 0 \leq r < 1.$

Thus

$$f(z) = \sum_{n=0}^{\infty} a_n z^n, \qquad |z| < 1.$$

Suppose f is holomorphic in D and defined and continuous on the closure: $\{z \mid |z| \leq 1\}$. Then the functions $a_n(r) = a_n r^n$ are also continuous at $r = 1$, and $a_n = a_n(1)$ is the nth Fourier coefficient of g_1. It follows that g_r is a convolution:

(5) $g_r = Q_r * g_1,$

where Q_r is the periodic distribution with Fourier coefficients $b_n = r^n$, $n > 0$ and $b_n = 0$, $n < 0$. Then

$$Q_r(\theta) = \sum_{n=0}^{\infty} r^n e^{in\theta} = \sum_{n=0}^{\infty} (re^{i\theta})^n$$

or

(6) $Q_r(\theta) = (1 - re^{i\theta})^{-1}.$

Equation (5) can be written

$$f(re^{i\theta}) = \frac{1}{2\pi} \int_0^{2\pi} (1 - re^{i(\theta-t)})^{-1} f(e^{it}) \, dt$$

$$= \frac{1}{2\pi i} \int_0^{2\pi} (e^{it} - re^{i\theta})^{-1} f(e^{it}) i e^{it} \, dt.$$

Setting $w = re^{i\theta}$ and $z = e^{it}$, we recover the Cauchy integral formula

(7)
$$f(w) = \frac{1}{2\pi i} \int_C f(z)(z - w)^{-1} \, dz,$$

where C is the unit circle. Thus (5) may be regarded as a version of the Cauchy integral formula.

Note that Q_r is a smooth periodic function when $0 \leq r < 1$. Therefore (5) defines a function in the disc if g_1 is only assumed to be a periodic distribution. In terms of the Fourier coefficients, if g_1 has Fourier coefficients $(a_n)_{-\infty}^{\infty}$ then those of g_r are $(a_n(r))_{-\infty}^{\infty}$ where $a_n(r) = 0$, $n < 0$, $a_n(r) = a_n r^n$, $n \geq 0$. These observations and the results of §§1 and 2 of Chapter 5 leads to the following theorem.

Theorem 7.1. *Suppose F is a periodic distribution with Fourier coefficients $(a_n)_{-\infty}^{\infty}$, where $a_n = 0$ for $n < 0$. Let f be the function defined in the unit disc by*

(8)
$$f(re^{i\theta}) = (F * Q_r)(\theta),$$

with Q_r given by (6). Then f is holomorphic in the unit disc and

(9)
$$f(z) = \sum_{n=0}^{\infty} a_n z^n, \qquad |z| < 1.$$

Moreover, F is the boundary value of f, in the sense that the distributions F_r defined by the functions $f_r(\theta) = f(re^{i\theta})$ converge to F in the sense of $\mathscr{P}'$ as $r \to 1$.

Conversely, suppose f is holomorphic in the unit disc. Then f is given by a convergent power series (9). If the sequence $(a_n)_0^{\infty}$ is of slow growth, i.e., if there are constants c, r such that

(10)
$$|a_n| \leq cn^r, \qquad n > 0,$$

then there is a distribution F such that (8) holds. If we require that the Fourier coefficients of F with negative indices vanish, then F is unique and is the boundary value of f in the sense above.

Condition (10) is not necessarily satisfied by the coefficients of a power series (9) converging in the disc. An example is

$$a_n = n^{\sqrt{n}}, \qquad n > 0.$$

Thus the condition (10) specifies a subset of the set of all holomorphic functions in the disc. This set of holomorphic functions is a vector space. Theorem 7.1 shows that this space corresponds naturally to the subspace of $\mathscr{P}'$ consisting of distributions whose negative Fourier coefficients all vanish.

Recall that $F \in \mathscr{P}'$ is in the Hilbert space L^2 if and only if its Fourier coefficients $(a_n)_{-\infty}^{\infty}$ satisfy

(11)
$$\sum |a_n|^2 < \infty.$$

For such distributions there is a result exactly like the preceding theorem.

Theorem 7.2. *Suppose F is a periodic distribution with Fourier coefficients $a_n = 0$ for $n < 0$. If $F \in \mathbf{L}^2$, then F is the boundary value of a function f,*

(12) $$f(z) = \sum_{n=0}^{\infty} a_n z^n, \qquad |z| < 1$$

with

(13) $$\sum_{n=0}^{\infty} |a_n|^2 < \infty.$$

The functions $f_r(\theta) = f(re^{i\theta})$ converge to F in the sense of $\mathbf{L}^2$ as $r \to 1$, and

$$\|F\|^2 = \sup_{0 \leq r < 1} \frac{1}{2\pi} \int_0^{2\pi} |f(re^{i\theta})|^2 \, d\theta.$$

Conversely, suppose f is defined in the unit disc by (12). Suppose either that (13) is true or that

(14) $$\sup_{0 \leq r < 1} \int_0^{2\pi} |f(re^{i\theta})|^2 \, d\theta < \infty.$$

Then both (13) and (14) hold, and the boundary value of f is a distribution $F \in \mathbf{L}^2$.

Proof. The first part of the theorem follows from Theorem 7.1 and the fact that (11) is a necessary condition for F to be in $\mathbf{L}^2$. The second part of the theorem is based on the identity

(15) $$\frac{1}{2\pi} \int_0^{2\pi} |f(re^{i\theta})|^2 \, d\theta = \sum_{n=0}^{\infty} |a_n|^2 r^{2n}, \qquad 0 \leq r < 1,$$

which is true because the Fourier coefficients of f_r are $a_n r^n$ for $n \geq 0$ and zero for $n < 0$. If (13) is true then (14) follows. Conversely, if (13) is false, then (15) shows that the integrals in (14) will increase to ∞ as $r \to 1$. Thus (13) and (14) are equivalent. By Theorem 7.1, if (13) holds then f has a distribution F as boundary value. The a_n are the Fourier coefficients of F, so (13) implies $F \in \mathbf{L}^2$. ☐

The set of holomorphic functions in the disc which satisfy (14) is a vector space which can be identified with the closed subspace of $\mathbf{L}^2$ consisting of distributions whose negative Fourier coefficients are all zero. Looked at either way this is a Hilbert space, usually denoted by

$$H^2 \quad \text{or} \quad H^2(D).$$

Exercises

1. Verify that

$$f(z) = \sum_{n=1}^{\infty} n^{\sqrt{n}} z^n$$

converges for $|z| < 1$ but that the coefficients do not satisfy (10).

2. Suppose f is holomorphic in the punctured disc $0 < |z| < 1$. Carry out the analysis of this section for

$$g(r, \theta) = f(re^{i\theta})$$

to deduce:

(a) $f(z) = \sum_{n=-\infty}^{\infty} a_n z^n$, $0 < |z| < 1$,
(b) If $|f(z)| \le M|z|^{-m}$ for some M, m, then

$$f(z) = \sum_{n=-m}^{\infty} a_n z^n.$$

Chapter 7

The Laplace Transform

§1. Introduction

It is useful to be able to express a given function as a sum of functions of some specified type, for example as a sum of exponential functions. We have done this for smooth periodic functions: if $u \in \mathscr{P}$, then

(1)
$$u(x) = \sum_{-\infty}^{\infty} a_n e^{inx},$$

where

(2)
$$a_n = \frac{1}{2\pi} \int_0^{2\pi} u(x)e^{-inx}\, dx.$$

Of course the particular exponential functions which occur here are precisely those which are periodic (period 2π). If $u: \mathbb{R} \to \mathbb{C}$ is a function which is not periodic, then there is no such natural way to single out a *sequence* of exponential functions for a representation like (1). One might suspect that (1) would be replaced by a continuous sum, i.e., an integral. This suspicion is correct. To derive an appropriate formula we start with the analogue of (2). Let

(3)
$$g(z) = \int_{-\infty}^{\infty} u(t)e^{-zt}\, dt,$$

when the integral exists. (Of course it may not exist for any $z \in \mathbb{C}$ unless restrictions are placed on u.)

If we are interested in functions u defined only on the half-line $[0, \infty)$, we may extend such a function to be zero on $(-\infty, 0]$. Then (3) for the extended function is equivalent to

(4)
$$g(z) = \int_0^{\infty} u(t)e^{-zt}\, dt.$$

If u is bounded and continuous, then the integral (4) will exist for each $z \in \mathbb{C}$ which has positive real part. More generally, if $a \in \mathbb{R}$ and $e^{-at}u(t)$ is continuous and bounded for $t > 0$, then the integral (4) exists for $\operatorname{Re} z > a$. Moreover, the function

(5)
$$g = Lu$$

is holomorphic in this half plane:

$$g'(z) = -\int_0^{\infty} tu(t)e^{-zt}\, dt, \qquad \operatorname{Re} z > a.$$

190

In particular, let

$$u_w(t) = e^{wt}, \qquad t > 0;$$
$$= 0, \qquad t \le 0, \qquad w \in \mathbb{C}.$$

Then for Re $z >$ Re w,

$$Lu_w(z) = \int_0^\infty e^{(w-z)t}\, dt = (z - w)^{-1}.$$

The operator L defined by (3) or (4) and (5) assigns, to certain functions on $\mathbb{R}$, functions holomorphic in half-planes in $\mathbb{C}$. This operator is clearly linear. We would like to *invert* it: given $g = Lu$, find u. Let us proceed formally, with no attention to convergence. Since Lu is holomorphic in some half-plane Re $z > a$, it is natural to invoke the Cauchy integral formula. Given z with Re $z > a$, choose b such that

$$a < b < \text{Re } z.$$

Let C be the vertical line Re $w = b$, traced in the upward direction. We consider C as "enclosing" the half-plane Re $w > b$, though traced in the wrong direction. A purely formal application of the Cauchy integral formula then gives

$$g(z) = Lu(z) = -\frac{1}{2\pi i}\int_C g(w)(w - z)^{-1}\, dw$$

$$= \frac{1}{2\pi i}\int_C g(w)Lu_w(z)\, dw.$$

If L has an inverse, then L^{-1} is also linear. Then we might expect to be able to interchange L^{-1} and integration in the preceding expression, to get

(6) $$u(t) = \frac{1}{2\pi i}\int_C g(w)e^{wt}\, dw,$$

or

$$u(t) = \frac{1}{2\pi}\int_{-\infty}^\infty g(b + is)e^{(b+is)t}\, ds.$$

Thus (3) or (4) and (6) are our analogues of (2) and (1) for periodic functions.

It is convenient for applications to interpret (3) and (6) for an appropriate class of *distributions F*. Thus if F is a continuous linear functional on a suitable space $\mathscr{L}$ of functions, we interpret (3) as

$$LF(z) = F(e_z), \qquad e_z(t) = e^{-zt}.$$

The space $\mathscr{L}$ will be chosen in such a way that each such continuous linear functional F can be extended to act on all the functions e_z for z in some half-plane Re $z > a$. Then the function LF will be holomorphic in this half-plane. We shall characterize those functions g such that $g = LF$ for some F, and give an appropriate version of the inversion formula (6).

The operator L is called the Laplace transform. It is particularly useful in connection with ordinary differential equations. To see why this might be so, let $u' = Du$ be the derivative of a function u. Substitution of u' for u in (3) and integration by parts (formally) yield

$$(7) \qquad\qquad [Lu'](z) = zLu(z).$$

More generally, suppose p is a polynomial

$$p(z) = a_m z^m + a_{m-1} z^{m-1} + \cdots + a_1 z + a_0.$$

Let $p(D)$ denote the corresponding *operator*

$$p(D) = a_m D^m + a_{m-1} D^{m-1} + \cdots + a_1 D + a_0,$$

i.e.,

$$p(D)u(z) = a_m u^{(m)}(x) + \cdots + a_1 u'(x) + a_0 u(x).$$

Then formally

$$(8) \qquad\qquad [Lp(D)u](z) = p(z)Lu(z).$$

Thus to solve the differential equation

$$p(D)u = v$$

we want

$$p(z)Lu(z) = Lv(z).$$

From (6), this becomes

$$(9) \qquad\qquad u(t) = \frac{1}{2\pi i} \int_C e^{zt} p(z)^{-1} Lv(z) \, dz.$$

As we shall see, all these purely formal manipulations can be justified.

Exercises

1. Show that the inversion formula (6) is valid for the functions u_w, i.e., if $a > \operatorname{Re} w$ and $C = \{z \mid \operatorname{Re} z = a\}$ then

$$\frac{1}{2\pi i} \int_C (z - w)^{-1} e^{zt} \, dz = e^{wt}, \qquad t > 0;$$

$$= 0, \qquad t < 0.$$

2. Suppose $u: [0, \infty) \to \mathbb{C}$ is bounded and continuous and suppose the derivative u' exists and is bounded and continuous on $(0, \infty)$. Show that

$$\int_0^\infty e^{-zt} u'(t) \, dt = z \int_0^\infty e^{-zt} u(t) \, dt - u(0),$$

$\operatorname{Re} z > 0$. Does this conflict with (7)?

§2. The space $\mathscr{L}$

Recall that a function $u\colon \mathbb{R} \to \mathbb{C}$ is said to be *smooth* if each derivative $D^k u$ exists and is continuous at each point of $\mathbb{R}$, $k = 0, 1, 2, \ldots$. In this section we shall be concerned with smooth functions u which have the property that each derivative of u approaches 0 very rapidly to the right. To be precise, let $\mathscr{L}$ be the set of all smooth functions $u\colon \mathbb{R} \to \mathbb{C}$ such that for every integer $k \geq 0$ and every $a \in \mathbb{R}$,

$$(1) \qquad \lim_{t \to +\infty} e^{at} D^k u(t) = 0.$$

This is equivalent to the requirement that for each integer $k > 0$, each $a \in \mathbb{R}$, and each $M \in \mathbb{R}$ the function

$$(2) \qquad e^{at} D^k u(t)$$

is bounded on the interval $[M, \infty)$. In fact (1) implies that (2) is bounded on every such interval. Conversely if (2) is bounded on $[M, \infty)$ then (1) holds when a is replaced by any smaller number $a' < a$.

It follows that if $u \in \mathscr{L}$, then for each k, a, M we have that

$$(3) \qquad |u|_{k,a,M} = \sup \{ e^{at} |D^j u(t)| \mid t \in [M, \infty), \, 0 \leq j \leq k \}$$

is finite. Conversely, if (3) is finite for every integer $k \geq 0$ and every $a \in \mathbb{R}$, $M \in \mathbb{R}$, then $u \in \mathscr{L}$.

The set of functions $\mathscr{L}$ is a vector space: it is easily checked that if u, $v \in \mathscr{L}$ and $b \in \mathbb{C}$ then bu and $u + v$ are in $\mathscr{L}$. Moreover,

$$(4) \qquad |bu|_{k,a,M} = |b| \, |u|_{k,a,M},$$

$$(5) \qquad |u + v|_{k,a,M} \leq |u|_{k,a,M} + |v|_{k,a,M}.$$

A sequence of functions $(u_n)_1^\infty \subset \mathscr{L}$ is said to *converge to $u \in \mathscr{L}$ in the sense of $\mathscr{L}$* if for each k, a, M,

$$|u_n - u|_{k,a,M} \to 0 \quad \text{as} \quad n \to \infty.$$

If so, we write

$$u_n \to u \ (\mathscr{L}).$$

The sequence $(u_n)_1^\infty \subset \mathscr{L}$ is said to be *a Cauchy sequence in the sense of $\mathscr{L}$* if for each k, a, M,

$$|u_n - u_m|_{k,a,M} \to 0 \quad \text{as} \quad m, n \to \infty.$$

As usual, a convergent sequence in this sense is a Cauchy sequence in this sense. The converse is also true.

Theorem 2.1. *$\mathscr{L}$ is a vector space. It is complete with respect to convergence as defined by the expressions (3); i.e., if $(u_n)_1^\infty \subset \mathscr{L}$ is a Cauchy sequence in the sense of $\mathscr{L}$, then there is a unique $u \in \mathscr{L}$ such that $u_n \to u \ (\mathscr{L})$.*

Proof. Let $(u_n)_1^\infty$ be a Cauchy sequence in the sense of $\mathscr{L}$. Taking (3) with $a = 0$, we see that each sequence of derivatives $(D^k u_n)_1^\infty$ is a uniform Cauchy sequence on each interval $[M, \infty)$. It follows, by Theorem 4.1 of Chapter 2, that there is a unique smooth function u such that $D^k u_n \to D^k u$ uniformly on each $[M, \infty)$. Now let a be arbitrary. Since $(e^{at} D^k u_n)_1^\infty$ is also a uniform Cauchy sequence on $[M, \infty)$ it follows that this sequence converges uniformly to $e^{at} D^k u$. Thus $u \in \mathscr{L}$ and $u_n \to u$ ($\mathscr{L}$). $\square$

It follows immediately from the definition of $\mathscr{L}$ that certain operations on functions in $\mathscr{L}$ give functions in $\mathscr{L}$. In particular, this is true of *differentiation*:

$$D^k u \in \mathscr{L} \qquad \text{if } u \in \mathscr{L};$$

translation:

$$T_s u \in \mathscr{L} \qquad \text{if } u \in \mathscr{L}$$

where

$$(T_s u)(t) = u(t - s);$$

and *complex conjugation*:

$$u^* \in \mathscr{L} \qquad \text{if } u \in \mathscr{L}$$

where

$$u^*(t) = u(t)^*.$$

It follows that if $u \in \mathscr{L}$, so are the *real and imaginary parts*:

$$\text{Re } u = \tfrac{1}{2}(u + u^*),$$

$$\text{Im } u = \frac{i}{2}(u^* - u).$$

Moreover, if $u \in \mathscr{L}$, so is the *integral* of u taken from the right:

$$S_+ u(t) = -\int_t^\infty u(s)\, ds.$$

In fact, $DS_+ u = u$ so

(6) $$|S_+ u|_{k,a,M} = |u|_{k-1,a,M}, \qquad k \geq 1.$$

For $k = 0$, note that for $t \geq M$, $a > 0$,

$$|S_+ u(t)| \leq \int_t^\infty |u(s)|\, ds \leq |u|_{0,a,M} \int_t^\infty e^{-as}\, ds$$

$$= a^{-1}|u|_{0,a,M} e^{-at}.$$

Thus

(7) $$|S_+ u|_{0,a,M} \leq a^{-1}|u|_{0,a,M}, \qquad a > 0.$$

The finiteness of

$$|S_+ u|_{0,a,M}$$

for $a \leq 0$ follows from finiteness for any $a > 0$. Thus $S_+ u \in \mathscr{L}$.

Lemma 2.2. *The operations of differentiation, translation, complex conjugation, and integration are continuous from $\mathscr{L}$ to $\mathscr{L}$ with respect to convergence in the sense of $\mathscr{L}$. Moreover, if $u \in \mathscr{L}$ then the difference quotient*

$$s^{-1}(T_{-s}u - u) \to Du \; (\mathscr{L})$$

as $s \to 0$.

Proof. These statements chiefly involve routine verifications. We shall prove the final statement. Given an integer $k \geq 0$, let $v = D^k u$. Suppose $t \geq M$ and $0 < |s| \leq 1$. The Mean Value Theorem implies

$$s^{-1}[T_{-s}v(t) - v(t)] = Dv(r)$$

where $|t - r| < |s|$. Then

$$D^k\{s^{-1}[T_{-s}u(t) - u(t)]\} - D^k Du(t) = Dv(r) - Dv(t)$$
$$= D^{k+1}u(r) - D^{k+1}u(t).$$

But

$$(8) \qquad |D^{k+1}u(r) - D^{k+1}u(t)| \leq c(a, M)e^{-at}, \qquad t \geq M.$$

The left side of (8) converges to zero as $s \to 0$, uniformly on bounded intervals. It follows that

$$|s^{-1}[T_{-s}u - u] - Du|_{k,a',M} \to 0$$

as $s \to 0$, for any $a' < a$. $\quad\square$

The functions e_z,

$$(9) \qquad e_z(t) = e^{-zt}, \qquad t \in \mathbb{R}, \qquad z \in \mathbb{C}$$

are not in $\mathscr{L}$ for any $z \in \mathbb{C}$. However, they may be approximated by functions from $\mathscr{L}$ in a suitable sense.

Lemma 2.3. *Suppose* $\mathrm{Re}\, z > a$. *There is a sequence* $(u_n)_1^\infty \subset \mathscr{L}$ *such that*

$$|u_n - e_z|_{k,a,M} \to 0 \quad \text{as} \quad n \to \infty$$

for each integer $k \geq 0$ and each $M \in \mathbb{R}$.

Proof. Choose a smooth function φ such that $\varphi(t) = 1$ if $t \leq 1$ and $\varphi(t) = 0$ if $t \geq 2$. (The existence of such a function is proved in §8 of Chapter 2.) Let

$$\varphi_n(t) = \varphi(t/n), \qquad u_n = \varphi_n e_z.$$

Then u_n is smooth and vanishes for $t \geq 2n$, so $u_n \in \mathscr{L}$. We shall consider in detail only the (typical) case $k = 1$.

$$e^{at}(Du_n(t) - De_z(t)) = e^{at}(\varphi_n(t)De_z(t) + e_z(t)D\varphi_n(t) - De_z(t))$$
$$= (1 - \varphi_n(t))ze^{(a-z)t} + D\varphi_n(t)e^{(a-z)t}.$$

Now both $1 - \varphi_n(t)$ and $D\varphi_n(t)$ are bounded independent of t and n, and vanish except on the interval $[n, +\infty)$. Therefore

$$|e^{at}(Du_n(t) - De_z(t))| \le c \exp(na - n \operatorname{Re} z),$$

c independent of n and t. Thus

$$|u_n - e_z|_{1,a,M} \to 0 \quad \text{as} \quad n \to \infty,$$

all M. The argument for other values of k is similar. $\square$

The following lemma relates the $|u|_{k,a,M}$ for different values of the indices k, a, M.

Lemma 2.4.   *Suppose k, k' are integers, and*

$$0 \le k \le k', \qquad a \le a', \qquad M \ge M'.$$

Suppose also either that $k = k'$ or that $a' > 0$. Then there is a constant c such that

(10) $$|u|_{k,a,M} \le c|u|_{k',a',M'}, \qquad \text{all } u \in \mathscr{L}.$$

Proof. It is sufficient to prove (10) in all cases when two of the three indices are the same. The case $k = k'$, $a = a'$ is trivial. The case $k = k'$, $M = M'$ is straightforward. Thus, suppose $a = a' > 0$ and $M = M'$. Let $k' = k + j$ and set $v = D^k u$. We may obtain $D^k u$ from v by repeated integrations:

$$D^k u = (S_+)^j v.$$

We use (7) repeatedly to get

$$\begin{aligned}
|u|_{k,a,M} = \;|D^k u|_{0,a,M} &\le a^{-j}|v|_{0,a,M} \\
&= a^{-j}|D^{k'} u|_{0,a,M} \\
&= a^{-j}|u|_{k',a,M}.
\end{aligned}$$ $\square$

If $u \in \mathscr{L}$, we set

$$|u|_k = |u|_{k,k,-k} = \sup\{|e^{kt}D^k u(t)| \mid t \ge -k\}.$$

Then the following is an easy consequence of Lemma 2.4.

Corollary 2.5.   *Suppose $(u_n)_1^\infty \subset \mathscr{L}$. Then*

$$u_n \to u \quad (\mathscr{L})$$

if and only if for each integer $k \ge 0$,

$$|u_n - u|_k \to 0 \quad \text{as} \quad n \to \infty.$$

Exercises

1. Show that $u(t) = \exp(-t^2)$ is in $\mathscr{L}$.
2. Show that if u, $v \in \mathscr{L}$, then the product uv is in $\mathscr{L}$.

3. Show that $u \in \mathcal{L}$, $z \in \mathbb{C}$ implies $e_z u \in \mathcal{L}$.

4. Complete the proof of Lemma 2.2.

§3. The space $\mathcal{L}'$

A linear functional on the vector space $\mathcal{L}$ is a function $F: \mathcal{L} \to \mathbb{C}$ such that

$$F(au) = aF(u), \qquad F(u + v) = F(u) + F(v).$$

A linear functional F on $\mathcal{L}$ is said to be *continuous* if

$$u_n \to u \ (\mathcal{L})$$

implies

$$F(u_n) \to F(u).$$

The set of all continuous linear functionals on $\mathcal{L}$ will be denoted $\mathcal{L}'$. An element $F \in \mathcal{L}'$ will be called a *distribution of type $\mathcal{L}'$*, or simply a *distribution*. An example is the *δ-distribution* defined by

$$(1) \qquad \delta(u) = u(0).$$

A second class of examples is given by

$$(2) \qquad F(u) = \int_0^\infty e^{zt} u(t) \, dt,$$

$z \in \mathbb{C}$.

Suppose $f: \mathbb{R} \to \mathbb{C}$ is a continuous function such that for some $a \in \mathbb{R}$, $M \in \mathbb{R}$,

$$(3) \qquad f(t) = 0, \qquad t \le M,$$

$$(4) \qquad e^{-at} f(t) \quad \text{is bounded.}$$

We may define

$$F_f: \mathcal{L} \to \mathbb{C}$$

by

$$(5) \qquad F_f(u) = \int_{-\infty}^\infty f(t) u(t) \, dt.$$

In fact the integrand is continuous and vanishes for $t \le M$. If $a' > a$ then on $[M, \infty)$ we have

$$|f(t)u(t)| = |e^{-at}f(t)| \, |e^{(a-a')t}| \, |e^{a't}u(t)|$$
$$\le c|u|_{0,a',M} e^{(a-a')t},$$

where c is a bound for $|e^{-at}f(t)|$. Therefore the integral (5) exists and

$$(6) \qquad |F_f(u)| \le c|u|_{0,a',M} \int_M^\infty e^{(a-a')t} \, dt.$$

It follows from (6) that F_f is continuous on $\mathscr{L}$, i.e., $F_f \in \mathscr{L}'$.

We say that $F \in \mathscr{L}'$ *is a function*, or *is defined by a function*, if there is a continuous function f satisfying (3) and (4), such that $F = F_f$.

Suppose $F \in \mathscr{L}'$ is defined by f. The translates $T_s f$ and the complex conjugate function f^* also define distributions. It is easy to check that if $g = T_s f$, i.e., $g(t) = f(t - s)$, $t \in \mathbb{R}$, then

$$F_g(u) = F_f(T_{-s}u), \qquad u \in \mathscr{L}.$$

Similarly,

$$F_{f*}(u) = F_f(u^*)^*, \qquad u \in \mathscr{L}.$$

We shall *define* the translates and the complex conjugate of any arbitrary $F \in \mathscr{L}'$ by

(7) $$(T_s F)(u) = F(T_{-s}u), \qquad u \in \mathscr{L}.$$

(8) $$F^*(u) = F(u^*)^*, \qquad u \in \mathscr{L}.$$

Similarly, the *real* and *imaginary* parts of $F \in \mathscr{L}'$ are *defined* by

(9) $$\mathrm{Re}\, F = \tfrac{1}{2}(F + F^*),$$

(10) $$\mathrm{Im}\, F = \frac{i}{2}(F^* - F).$$

We say that F is *real* if $F = F^*$.

Suppose $F \in \mathscr{L}'$ is defined by f and suppose the derivative Df exists, is continuous, and satisfies (3) and (4). Integration by parts gives

$$F_{Df}(u) = -F(Du), \qquad u \in \mathscr{L}.$$

Therefore we shall *define* the derivative DF of an arbitrary $F \in \mathscr{L}'$ by

(11) $$DF(u) = -F(Du), \qquad u \in \mathscr{L}.$$

Generally, for any integer $k \geq 0$ we define

(12) $$D^k F(u) = (-1)^k F(D^k u).$$

Proposition 3.1. *The set $\mathscr{L}'$ is a vector space. If $F \in \mathscr{L}'$, then the translates $T_s F$, the complex conjugate F^*, and the derivatives $D^k F$ are in $\mathscr{L}'$.*

Proof. All these statements follow easily from the definitions and the continuity of the operations in $\mathscr{L}$; see Lemma 2.2. For example, $D^k F: \mathscr{L} \to \mathbb{C}$ is certainly linear. If $u_n \to u$ $(\mathscr{L})$, then

$$D^k u_n \to D^k u \quad (\mathscr{L}),$$

so

$$D^k F(u_n) = (-1)^k F(D^k u_n) \to (-1)^k F(D^k u) = D^k F(u).$$

Thus $D^k F$ is continuous. □

A sequence $(F_n)_1^\infty \subset \mathscr{L}'$ is said to *converge to* $F \in \mathscr{L}'$ *in the sense of* $\mathscr{L}'$ if for each $u \in \mathscr{L}$

$$F_n(u) \to F(u) \quad \text{as} \quad n \to \infty.$$

We denote this by

$$F_n \to F \ (\mathscr{L}').$$

The operations defined above are continuous with respect to this notion of convergence.

Proposition 3.2. *Suppose*

$$F_n \to F \ (\mathscr{L}'), \qquad G_n \to G \ (\mathscr{L}'),$$

and suppose $a \in \mathbb{C}$, $s \in \mathbb{R}$. *Then*

$$\begin{aligned} aF_n &\to aF \ (\mathscr{L}'), \\ F_n + G_n &\to F + G \ (\mathscr{L}'), \\ T_s F_n &\to T_s F \ (\mathscr{L}'), \\ D^k F_n &\to D^k F \ (\mathscr{L}'). \end{aligned}$$

Moreover, the difference quotient

$$s^{-1}[T_{-s}F - F] \to DF \, (\mathscr{L}')$$

as $s \to 0$.

Proof. All except the last statement follow immediately from the definitions. To prove the last statement we use Lemma 2.2:

$$\begin{aligned} s^{-1}[T_{-s}F - F](u) &= F(s^{-1}[T_s u - u]) \\ &\to F(-Du) = (DF)(u). \qquad \square \end{aligned}$$

The following theorem gives a very useful necessary and sufficient condition for a linear functional on $\mathscr{L}$ to be continuous.

Theorem 3.3. *Suppose* $F: \mathscr{L} \to \mathbb{C}$ *is linear. Then* F *is continuous if and only if there are an integer* $k \geq 0$ *and constants* a, M, $K \in \mathbb{R}$ *such that*

(13) $$|F(u)| \leq K|u|_{k,a,M}, \quad \text{all } u \in \mathscr{L}.$$

Proof. Suppose (13) is true. If $u_n \to u \ (\mathscr{L})$ then

$$|F(u_n) - F(u)| \leq K|u_n - u|_{k,a,M} \to 0.$$

Thus F is continuous.

To prove the converse, suppose that (13) is *not* true for any k, a, M, K. In particular, for each positive integer k we may find a $v_k \in \mathscr{L}$ such that

$$|F(v_k)| \geq k|v_k|_k = k|v_k|_{k,k,-k} \neq 0.$$

Let $u_k \in \mathscr{L}$ be

$$u_k = k^{-1}|v_k|_k^{-1}v_k.$$

Then

$$|u_k|_k = k^{-1}, \qquad |F(u_k)| \geq 1.$$

But Corollary 2.5 implies that

$$u_k \to 0 \quad (\mathcal{L}).$$

Since $F(u_k)$ does not converge to 0, F is not continuous. □

The *support* of a function $u\colon \mathbb{R} \to \mathbb{C}$ is defined as the smallest closed subset A of $\mathbb{R}$ such that $u(t) = 0$ for every $t \notin A$. Another way of phrasing this is that t is *not* in the support of u if and only if there is an $\varepsilon > 0$ such that u is zero on the interval $(t - \varepsilon, t + \varepsilon)$. The support of u is denoted

$$\operatorname{supp}(u).$$

Condition (3) on a function f can be written

$$\operatorname{supp}(f) \subset [M, \infty).$$

The *support* of a *distribution* $F \in \mathcal{L}'$ can be defined similarly. A point $t \in \mathbb{R}$ is *not* in the support of F if and only if there is an $\varepsilon > 0$ such that

$$F(u) = 0$$

whenever $u \in \mathcal{L}$ and $\operatorname{supp}(u) \subset (t - \varepsilon, t + \varepsilon)$. We denote the support of F also by

$$\operatorname{supp}(F).$$

Theorem 3.3 implies that any $F \in \mathcal{L}'$ has support in a half line.

Corollary 3.4. *If $F \in \mathcal{L}'$, there is $M \in \mathbb{R}$ such that*

$$\operatorname{supp}(F) \subset [M, \infty).$$

Proof. Choose k, a, M such that (13) is true for some K. If $u \in \mathcal{L}$ and

$$\operatorname{supp}(u) \subset (-\infty, M),$$

then

$$|u|_{k,a,M} = 0$$

so $F(u) = 0$. Therefore each $t < M$ is not in the support of F. □

Exercises

1. Compute the following in the case $F = \delta$:

$$T_s F(u), \qquad F^*(u), \qquad \operatorname{Re} F, \qquad \operatorname{Im} F, \qquad D^k F(u).$$

2. Show that if F is given by (2), then

$$DF = \delta + zF.$$

3. Suppose $F = D^j \delta$. For what constants k, a, M, K is (13) true?

4. Find supp $(D^k\delta)$.
5. Prove that if F is defined by a function f, then

$$\text{supp}\,(F) = \text{supp}\,(f).$$

§4. Characterization of distributions of type $\mathscr{L}'$

If $f: \mathbb{R} \to \mathbb{C}$ is a function satisfying the two conditions (3) and (4) of §3 and if k is an integer ≥ 0, then

(1) $$D^k F_f$$

is a distribution of type $\mathscr{L}'$. In this section we shall prove that, conversely, any $F \in \mathscr{L}'$ is of the form (1) for some k and some function f. The proof depends on two notions: the order of a distribution and the integral (from the left) of a distribution.

A distribution $F \in \mathscr{L}'$ is said to be *of order* k if (13) of §3 is true, i.e., if for some real constants a, M, and K,

(2) $$|F(u)| \leq K|u|_{k,a,M}, \qquad \text{all } u \in \mathscr{L}.$$

By Theorem 3.3, each $F \in \mathscr{L}'$ is of order k for *some* $k \geq 0$.

Suppose $F \in \mathscr{L}'$ is defined by a function f. Let g be the integral of f from the left:

$$g(t) = \int_{-\infty}^{t} f(s)\, ds = \int_{M}^{t} f(s)\, ds,$$

where supp $(f) \subset [M, \infty)$. If $u \in \mathscr{L}$ and $v = S_+ u$, then $Dv = u$. Integration by parts gives

$$F_g(u) = \int_{-\infty}^{\infty} g(t)v'(t)\, dt = -\int_{-\infty}^{\infty} g'(t)v(t)\, dt = -F_f(S_+ u).$$

For an arbitrary $F \in \mathscr{L}'$ we *define* the *integral of F* (from the left), denoted $S_- F$, by

$$S_- F(u) = -F(S_+ u).$$

Proposition 4.1. *If $F \in \mathscr{L}'$ then the integral $S_- F$ is in $\mathscr{L}'$ and*

(3) $$D(S_- F) = F = S_-(DF).$$

If F is of order $k \geq 1$, then $S_- F$ is of order $k - 1$.

Proof. Clearly $S_- F$ is linear. The continuity follows from the definition and the fact that S_+ is continuous in $\mathscr{L}$. The identity (3) is a matter of manipulation:

$$D(S_- F)(u) = -S_- F(Du) = F(S_+ Du) = F(u), \qquad u \in \mathscr{L},$$

and similarly for the other part. $\quad\square$

To prove that every $F \in \mathscr{L}'$ is of the form (1), we want to integrate F enough times to get a distribution defined by a function. To motivate this, we consider first a function $f: \mathbb{R} \to \mathbb{C}$ such that the first and second derivatives are continuous and such that

$$\text{supp}\,(f) \subset [M, \infty),$$

some M. Then by integrating twice and changing the order of integration (see §7 of Chapter 2) we get

$$f(t) = \int_{-\infty}^{t} Df(s)\, ds = \int_{-\infty}^{t} \int_{-\infty}^{s} D^2f(r)\, dr\, ds$$

$$= \int_{-\infty}^{t} \int_{r}^{t} D^2f(r)\, ds\, dr$$

$$= \int_{-\infty}^{t} (t - r)D^2f(r)\, dr.$$

Let

(4)
$$h(t) = |t|, \qquad t \le 0,$$
$$= 0, \qquad t > 0.$$

Then our equation is

$$f(t) = \int_{-\infty}^{\infty} h(r - t)D^2f(r)\, dr$$

or

(5)
$$f(t) = \int_{-\infty}^{\infty} D^2f(r)T_th(r)\, dr.$$

We would like to interpret (5) as the action of the distribution defined by D^2f on the function T_th; however T_th is not in $\mathscr{L}$. Nevertheless, h can be approximated by elements of $\mathscr{L}$.

Lemma 4.2. *Let h be defined by (4). There is a sequence $(h_n)_1^\infty \subset \mathscr{L}$ such that $\text{supp}(h_n) \subset (-\infty, 0]$ and*

(6)
$$|h_n - h|_{0,a,M} \to 0 \quad \text{as} \quad n \to \infty$$

for each $a \in \mathbb{R}$, $M \in \mathbb{R}$.

Proof. There is a smooth function $\varphi: \mathbb{R} \to \mathbb{R}$ such that $0 \le \varphi(t) \le 1$ for all t and $\varphi(t) = 1$, $t \le -2$, $\varphi(t) = 0$, $t \ge -1$; see §8 of Chapter 2. Let

$$h_n(t) = \varphi(nt)h(t) = \varphi_n(t)h(t).$$

Then h_n is smooth, since φ_n is zero in an interval around 0 and h is smooth except at 0. Also $h_n(t) = h(t)$ except in the interval $(-2/n, 0)$, and

$$|h_n(t) - h(t)| \le 2/n, \qquad t \in (-2/n, 0).$$

Thus $h_n \in \mathscr{L}$ and (6) is true.   ∎

Theorem 4.3. *Suppose $F \in \mathscr{L}'$ is of order $k - 2$, where k is an integer ≥ 2. Then there is a unique function f such that*

$$F = D^k(F_f).$$

Proof. Suppose first that $k = 2$, F is of order 0. Let h be the function defined by (4) and let $(h_n)_1^\infty \subset \mathscr{L}$ be as in Lemma 4.2. Choose $a, M, K \in \mathbb{R}$ such that

(7) $\qquad\qquad |F(u)| \leq K|u|_{0,a,M}, \qquad$ all $u \in \mathscr{L}.$

We may suppose $a > 0$. For each $s \in \mathbb{R}$ the translates $T_s h_n$ also converge to $T_s h$:

(8) $\qquad\qquad |T_s h_n - T_s h|_{0,a,M} \to 0 \quad$ as $\quad n \to \infty.$

It follows from (7) and (8) that for each $s \in \mathbb{R}$

$$F(T_s h_n) \quad \text{converges as} \quad n \to \infty.$$

Let $f(s)$ be the limit of this sequence. Then

(9) $\qquad\qquad |f(s)| \leq \lim_{n \to \infty} K|T_s h_n|_{0,a,M} = K|T_s h|_{0,a,M}.$

But

(10) $\qquad\qquad |T_s h|_{0,a,M} = 0 \qquad\qquad$ if $s \leq M,$

(11) $\qquad\qquad |T_s h|_{0,a,M} \leq (s - M)e^{as} \qquad$ if $s > M.$

Thus

$$\text{supp}\,(f) \subset [M, \infty)$$

and for any $a' > a$ there is a constant c such that

$$|f(s)| \leq ce^{a's}, \qquad \text{all } s \in \mathbb{R}.$$

If f is continuous, it follows that f defines a distribution $F_f \in \mathscr{L}'$. Suppose $s < t$. Taking limits we get

$$|f(t) - f(s)| \leq K|T_t h - T_s h|_{0,a,M}$$
$$\leq Ke^{a'}(t - s).$$

Thus f is continuous.

Let $f_n(s) = F(T_s h_n)$. Then $f_n(s) = 0$ if $s \leq M$. For $s > M$,

$$|f_n(s) - f(s)| \leq K|T_s h_n - T_s h|_{0,a,M}$$
$$\leq 2Ke^{as}/n.$$

Therefore if $u \in \mathscr{L}$,

$$F_f(u) = \int_{-\infty}^{\infty} f(s)u(s)\, ds = \lim G_n(u)$$

where G_n is the distribution defined by f_n. Then

$$D^2 G_n(u) = G_n(D^2 u) = \int_{-\infty}^{\infty} F(T_s h_n) D^2 u(s)\, ds.$$

Let v_n be the function defined by

$$v_n(t) = \int_{-\infty}^{\infty} T_s h_n(t) D^2 u(s)\, ds$$

$$= \int_t^{\infty} h_n(t - s) D^2 u(s)\, ds.$$

Since $D^2 u \in \mathscr{L}$, and $h_n(t - s) = 0$ if $s \le t$, the integral converges. Moreover, it is not difficult to see that the integral is the limit of its Riemann sums

$$v_{n,N}(t) = \frac{1}{N} \sum_{m=1-N^2}^{N^2} h_n(t - m/N) D^2 u(m/N),$$

in the sense that

$$|v_{n,N} - v_n|_{0,a,M} \to 0 \quad \text{as} \quad N \to \infty.$$

In fact, let

$$w_{n,N}(t) = \int_{-N}^{N} h_n(t - s) D^2 u(s)\, ds.$$

Then $|v_n - w_{n,N}|_{0,a,M} \to 0$ as $N \to \infty$, while for $t \ge M$

$$|w_{n,N}(t) - v_{n,N}(t)|$$

$$= \left| \sum_{1-N^2}^{N^2} \int_{(m-1)/N}^{m/N} \left[h_n(t - s) D^2 u(s) - h_n\left(t - \frac{m}{N}\right) D^2 u\left(\frac{m}{N}\right) \right] ds \right|$$

$$\le \frac{C}{N} \int_t^N e^{-as}\, ds < \frac{C}{aN} e^{-at}.$$

Since $h_n(t - s) = 0$ for $s \le t$. Therefore

$$F(v_n) = \lim_N F(v_{n,N})$$

$$= \lim_N \frac{1}{N} \sum_{m=N^2}^{N^2} f_n(m/N) D^2 u(m/N)$$

$$= \int_{-\infty}^{\infty} f_n(s) D^2 u(s)\, ds = D^2 G_n(u).$$

Now let

$$v(t) = \int_{-\infty}^{\infty} h(t - s) D^2 u(s)\, ds.$$

Then

$$|v_n - v|_{0,a,M} \to 0 \quad \text{as} \quad n \to \infty.$$

In fact, for $t \ge M$

$$|v_n(t) - v(t)| \le \int_t^{\infty} |h_n(t - s) - h(t - s)|\, |D^2 u(s)|\, ds$$

$$\le \int_0^{2/n} \frac{2}{n} |D^2 u(s + t)|\, ds \le c e^{-at}/n.$$

Therefore
$$D^2F_f(u) = F_f(D^2u) = \lim G_n(D^2u)$$
$$= \lim F(v_n) = F(v).$$

But
$$v(t) = \int_t^\infty (s - t)D^2u(s)\, ds$$
$$= -\int_t^\infty Du(s)\, ds = u(t).$$

Thus $D^2F_f = F$.

Now suppose F is of order $k - 2 > 0$. Let $G = (S_-)^{k-2}F$. Then G is of order 0, and by what we have just shown, there is an f such that
$$D^2F_f = G.$$

But then
$$D^kF_f = D^{k-2}G = F.$$

Finally, we must prove *uniqueness*. This is equivalent to showing that $DF = 0$ implies $F = 0$. But $F = S_-(DF)$, so this is the case. □

Exercises

1. Show that $D^k\delta$ is of order k but not of order $k - 1$.
2. Find the function f of Theorem 4.3 when $F = \delta$. Compute $DF_f = S_-\delta$.
3. Show that if $\operatorname{supp}(F) \subset [M, \infty)$, then $\operatorname{supp}(S_-F) \subset [M, \infty)$, and conversely.
4. Suppose $F \in \mathscr{L}'$ and the support of F consists of the single point 0. Show that F is of the form
$$\sum_{k=0}^m a_k D^k\delta,$$

where the a_k's are constants.

§5. Laplace transforms of functions

Suppose that f is a function which defines a distribution of type $\mathscr{L}'$, i.e., $f: \mathbb{R} \to \mathbb{C}$ is continuous, and

(1) $$\operatorname{supp}(f) \subset [M, \infty),$$

(2) $$|f(t)| \leq Ke^{at}, \quad \text{all } t.$$

If $z \in \mathbb{C}$, let e_z be as before:
$$e_z(t) = e^{-zt}, \quad t \in \mathbb{R}.$$

If $\operatorname{Re} z = b > a$ then
$$|f(t)e_z(t)| \leq Ke^{(a-b)t}.$$

Since $f(t) = 0$ for $t < M$, the integral

(3)
$$\int_{-\infty}^{\infty} f(t)e_z(t) \, dt = \int_M^{\infty} f(t)e_z(t) \, dt$$

exists when Re $z > a$. The *Laplace transform of the function f* is the function Lf defined by (3):

(4)
$$Lf(z) = \int_{-\infty}^{\infty} e^{-zt}f(t) \, dt, \qquad \text{Re } z > a.$$

Theorem 5.1. *Suppose $f: \mathbb{R} \to \mathbb{C}$ is continuous and satisfies* (1) *and* (2). *Then the Laplace transform Lf is holomorphic in the half plane*

$$\{z \mid \text{Re } z > a\}.$$

The derivative is

(5)
$$(Lf)'(z) = -\int_{-\infty}^{\infty} e^{-zt}tf(t) \, dt.$$

The Laplace transform satisfies the estimate

(6) $|Lf(z)| \leq K(\text{Re } z - a)^{-1} \exp(M(a - \text{Re } z)), \qquad \text{Re } z > a.$

Proof

$$(w - z)^{-1}[Lf(w) - Lf(z)] = \int_M^{\infty} g(w, z, t)f(t) \, dt$$

where

$$g(w, z, t) = (w - z)^{-1}[e^{-wt} - e^{-zt}].$$

Suppose Re z and Re w are $\geq b > a$. Let

$$h(s) = \exp[-(1 - s)z - sw]t, \qquad 0 \leq s \leq 1.$$

Then

$$g(w, z, t) = (w - z)^{-1}[h(1) - h(0)].$$

An application of the Mean Value Theorem to the real and imaginary parts of h shows that

$$|h(1) - h(0)| \leq ct|w - z|e^{-bt}, \qquad t \geq 0.$$

Thus as $w \to z$,

$$|g(w, z, t)f(t)| \leq c_1 e^{-\varepsilon t}, \qquad \text{where} \quad \varepsilon > 0.$$

Moreover,

$$g(w, z, t)f(t) \to -te^{-zt}f(t)$$

as $w \to z$, uniformly on each interval $[M, N]$. It follows that Lf is differentiable and that (5) is true.

The estimate (6) follows easily from (1) and (2):

$$|Lf(z)| \leq \int_M^\infty |f(t)e^{-zt}| \, dt$$

$$\leq K \int_M^\infty \exp t(a - \operatorname{Re} z) \, dt$$

$$= K(\operatorname{Re} z - a)^{-1} \exp (M(a - \operatorname{Re} z)). \qquad \square$$

We want next to invert the process: determine f, given Lf.

Theorem 5.2. *Suppose f satisfies the conditions of Theorem 5.1, and let $g = Lf$. Given $b > \max\{a, 0\}$, let C be the line*

$$\{z \mid \operatorname{Re} z = b\}.$$

Then f is the second derivative of the function F defined by

(7)
$$F(t) = \frac{1}{2\pi i} \int_C e^{zt} z^{-2} g(z) \, dz.$$

Proof. By (6), g is bounded on the line C. Therefore the integral (7) exists. Moreover, if

$$g_N(z) = \int_{-N}^N e^{-zt} f(t) \, dt$$

then the g_N are bounded uniformly on the line C and converge uniformly to g. Thus $F(t)$ is the limit as $N \to \infty$ of F_N, where

$$F_N(t) = \frac{1}{2\pi i} \int_C e^{zt} z^{-2} g_N(z) \, dz$$

$$= \frac{1}{2\pi i} \int_C e^{zt} z^{-2} \int_{-N}^N e^{-zs} f(s) \, ds \, dz$$

$$= \int_{-N}^N \left\{ \frac{1}{2\pi i} \int_C e^{z(t-s)} z^{-2} \, dz \right\} f(s) \, ds.$$

Let us consider the integral in braces. When $s > t$ the integrand is holomorphic to the right of C and has modulus $\leq k|z|^{-2}$ for some constant k. Let C_R be the curve consisting of the segment $\{\operatorname{Re} z = b \mid |z - b| \leq R\}$ and the semicircle $\{\operatorname{Re} z > b \mid |z - b| = R\}$. Then the integral of

$$e^{z(t-s)} z^{-2}$$

over C_R in the counterclockwise direction is zero, and the limit as $R \to \infty$ is

$$-\int_C e^{z(t-s)} z^{-2} \, dz.$$

Thus the integral in braces vanishes for $s > t$. When $s < t$, let C'_R be the reflection of C_R about the line C. Then for $R > b$,

$$\frac{1}{2\pi i} \int_{C'_R} e^{z(t-s)}z^{-2}\, dz = \frac{d}{dz} e^{z(t-s)}\bigg|_{z=0}$$

$$= t - s.$$

Taking the limit as $R \to \infty$ we get

$$\frac{1}{2\pi i} \int_C e^{z(t-s)}z^{-2}\, dz = t - s, \qquad s < t.$$

Thus

$$F(t) = \int_{-\infty}^{t} (t - s)f(s)\, ds.$$

It follows that $D^2 F = f$. $\square$

We can get a partial converse of Theorem 5.1.

Theorem 5.3. *Suppose g is holomorphic in the half plane*

$$\{z \mid \operatorname{Re} z > a\}$$

and satisfies the inequality

(8) $$|g(z)| \le K(1 + |z|)^{-2} \exp(-M \operatorname{Re} z).$$

Then there is a unique continuous function $f: \mathbb{R} \to \mathbb{C}$ with the properties

(9) $$\operatorname{supp} f \subset [M', \infty), \quad some \quad M',$$

(10) $$|f(t)| \le Ke^{bt}, \quad all \ t, for \ some \ b,$$

(11) $$Lf(z) = g(z) \quad for \quad \operatorname{Re} z > b.$$

Moreover, we may take $M' = M$ in (9) and any $b > a$ in (10) and (11).

Proof. Choose $b > a$ and let $C(b)$ be the line $\{z \mid \operatorname{Re} z = b\}$. Let

(12) $$f(t) = \frac{1}{2\pi i} \int_{C(b)} e^{zt}g(z)\, dz.$$

It follows from (8) that the integral exists and defines a continuous function. It follows from (8) and an elementary contour integration argument that (12) is independent of b, provided $b > a$. Moreover, (8) gives the estimate

(13) $$|f(t)| \le Ce^{b(t-M)}, \quad b > a,$$

where C is independent of b and t. This implies (10). If $t < M$ we may take $b \to +\infty$ in (13) and get

$$f(t) = 0, \ t < M.$$

Thus the Laplace transform of f can be defined for $\operatorname{Re} z > a$. If $\operatorname{Re} w > a$, choose

$$a < b < \operatorname{Re} w.$$

Then

$$Lf(w) = \int_M^\infty e^{-wt} \left\{ \frac{1}{2\pi i} \int_{C(b)} e^{zt} g(z)\, dz \right\} dt.$$

Since

$$|e^{-wt+zt}g(z)| \le c(1 + |z|)^{-2} \exp\left[-Mb + t(b - \operatorname{Re} w)\right]$$

for $t \in \mathbb{R}$ and $z \in C(b)$, we may interchange the order of integration. This gives

$$Lf(w) = \frac{1}{2\pi i} \int_{C(b)} g(z) \left\{ \int_M^\infty e^{(z-w)t}\, dt \right\} dz$$

$$= \frac{-1}{2\pi i} \int_{C(b)} g(z) e^{(z-w)M}(z - w)^{-1}\, dz.$$

In the half plane $\operatorname{Re} z > b$ we have

$$|g(z)e^{(z-w)M}| \le c(1 + |z|)^{-2}.$$

Therefore a contour integration argument and the Cauchy integral formula give

$$Lf(w) = g(z)e^{(z-w)M}\big|_{z=w}$$
$$= g(w).$$

Finally, we must show that f is unique. This is equivalent to showing that $Lf \equiv 0$ implies $f \equiv 0$. But this follows from Theorem 5.2.   □

Exercises

1. Let $f(t) = 0$, $t < 0$; $f(t) = te^{at}$, $t \ge 0$. Compute Lf and verify that

$$f(t) = \frac{1}{2\pi i} \int_C e^{zt} Lf(z)\, dz,$$

where C is an appropriate line.

2. Show that the Laplace transform of the translate of a function f satisfies

$$L(T_s f)(z) = e^{-zs} Lf(z).$$

3. Suppose both f and Df are functions satisfying (1) and (2). Show that

$$L(Df)(z) = zLf(z).$$

4. Compute Lf when

$$f(t) = 0, \qquad t < 0; \qquad f(t) = t^n, \qquad t > 0,$$

where n is a positive integer.

5. Suppose f satisfies (1) and (2), and let $g = e_w f$. Show that

$$Lg(z) = Lf(z + w).$$

6. Compute Lf when

$$f(t) = 0, \quad t < 0; \quad f(t) = e^{wt}t^n, \quad t > 0.$$

7. Compute Lf when $f(t) = 0, t < 0$;

$$f(t) = \int_0^t \sin s \, ds, \quad t > 0.$$

§6. Laplace transforms of distributions

Suppose $F \in \mathscr{L}'$. Theorem 3.3 states that there are constants k, a, M, K such that

(1) $$|F(u)| \le K|u|_{k,a,M}, \quad \text{all } u \in \mathscr{L}.$$

If Re $z > a$, then Lemma 2.3 states that there is a sequence $(u_n)_1^\infty \subset \mathscr{L}$ such that

(2) $$|u_n - e_z|_{k,a,M} \to 0 \quad \text{as} \quad n \to \infty,$$

where $e_z(t) = e^{-zt}$. Now (1) and (2) imply that $(F(u_n))_{n=1}^\infty$ is a Cauchy sequence in $\mathbb{C}$. We shall define the *Laplace transform LF* by

(3) $$LF(z) = \lim_{n \to \infty} F(u_n).$$

In view of (2) we shall write, symbolically,

(4) $$LF(z) = F(e_z)$$

even though $e_z \notin \mathscr{L}$. Note that if $(v_n)_1^\infty \subset \mathscr{L}$ and

$$|v_n - e_z|_{k,a,M} \to 0 \quad \text{as} \quad n \to \infty$$

then

$$|F(v_n) - F(u_n)| \le K|v_n - u_n|_{k,a,M} \to 0$$

as $n \to \infty$. Thus $LF(z)$ is independent of the particular sequence used to approximate e_z.

Proposition 6.1. *Suppose* $F, G \in \mathscr{L}'$ *and* $b \in \mathbb{C}$. *Then on the common domain of definition*

(5) $$L(bF) = bLF;$$

(6) $$L(F + G) = LF + LG;$$

(7) $$L(T_sF)(z) = e^{-zs}LF(z);$$

(8) $$L(D^kF)(z) = z^kLF(z);$$

(9) $$L(S_-F)(z) = z^{-1}LF(z), \quad z \ne 0.$$

Moreover,

(10) $$L(e_wF)(z) = LF(z + w),$$

where $e_w F$ is the distribution defined by

(11) $$e_w F(u) = F(e_w u), \qquad u \in \mathscr{L}.$$

If F is defined by a function f, then

(12) $$LF = Lf.$$

Proof. The identities (5) and (6) follow immediately from the definitions. If $(u_n)_1^\infty \subset \mathscr{L}$ satisfies (2) then the sequence $(T_{-s} u_n)_1^\infty$ approximates $T_{-s} e_z$ in the same sense. But

$$T_{-s} e_z = e^{-zs} e_z,$$

so

$$L(T_s F)(z) = \lim T_s F(u_n) = \lim F(T_{-s} u_n)$$
$$= e^{-zs} \lim F(u_n) = e^{-zs} LF(z).$$

This proves (7), and the proofs of (8), (9) and (10) are similar. Note that $u \in \mathscr{L}$ implies $e_w u \in \mathscr{L}$ and

$$u_n \to u \ (\mathscr{L})$$

implies

$$e_w u_n \to e_w u \ (\mathscr{L}).$$

Therefore (11) does define a distribution.

Finally, (12) follows from the definitions. □

We can now generalize Theorems 5.1 and 5.2 to distributions.

Theorem 6.2. *Suppose $F \in \mathscr{L}'$ and suppose F satisfies (1). Then the Laplace transform LF is holomorphic in the half plane*

$$\{z \mid \mathrm{Re}\ z > a\}.$$

Moreover,

(13) $$F = D^{k+2} F_f,$$

where f is the function defined by

(14) $$f(t) = \frac{1}{2\pi i} \int_C e^{zt} z^{-k-2} LF(z)\, dz.$$

Here C is the line $\{z \mid \mathrm{Re}\ z = b\}$, where $b > \max\{a, 0\}$.

Proof. We know by Theorem 4.3 that there is a function f such that

$$F = D^{k+2} F_f.$$

It was shown in the proof of Theorem 4.3 that for any $b > \max\{a, 0\}$ there is a constant c such that

$$|f(t)| \le ce^{bt}, \qquad \text{all } t.$$

Therefore Lf is holomorphic for Re $z >$ max $\{a, 0\}$. By Proposition 6.1,

(15) $LF(z) = z^{k+2}Lf(z), \qquad \text{Re } z > \text{max } \{a, 0\}.$

Therefore LF is holomorphic in this half plane. This completes the proof of the first statement in the case $a \geq 0$. When $a < 0$, let $G = e_a F$. Then (1) implies

$$|G(u)| \leq K|u|_{k,0,M}.$$

Thus by the argument just given, LG is holomorphic for Re $z > 0$. Since $LF(z) = LG(z - a)$, LF is holomorphic for Re $z > a$.

Now let C be the line $\{z \mid \text{Re } z = b\}$, where $b >$ max $\{a, 0\}$. Let f be the function such that (13) is true. Then by Theorem 5.2 and equation (15), f is the second derivative of the function

(16) $$g(t) = \frac{1}{2\pi i} \int_C e^{zt} z^{-k-4} LF(z) \, dz.$$

From the definition of LF it follows that on C

(17) $$|LF(z)| \leq K|e_z|_{k,a,M} = K|z|^k e^{aM-bM}.$$

Using (17) we may justify differentiating (16) twice under the integral sign to get (14). □

Theorem 6.2 implies, in particular, that if $LF \equiv 0$ then $F = 0$.

Given a holomorphic function g, how can one tell whether it is the Laplace transform of a distribution?

Theorem 6.3.   *Suppose g is holomorphic in a half plane*

$$\{z \mid \text{Re } z > a\}.$$

Then g is the Laplace transform of a distribution $F \in \mathcal{L}'$ if and only if there are constants k, a, M, K_1 such that

(18) $$|g(z)| \leq K_1(1 + |z|)^k \exp(-M \text{ Re } z), \qquad \text{Re } z > a.$$

Proof.   Suppose $g = LF$, where $F \in \mathcal{L}'$. Then there are k, a, M, K such that (1) is true. Then Re $z > a$ implies

$$|LF(z)| \leq K|e_z|_{k,a,M} \leq K_1(1 + |z|)^k \exp(-M \text{ Re } z),$$

where $K_1 = Ke^{aM}$.

Conversely, suppose (18) is true. Take $b >$ max $\{a, 0\}$. We may apply Theorem 5.3 to

$$h(z) = z^{-k-2}g(z)$$

to conclude that

$$h = Lf,$$

where f is continuous,

$$\text{supp} f \subset [M, \infty),$$
$$|f(t)| \leq ce^{bt}.$$

Let $F = D^{k+2}F_f$. Then

$$LF(z) = z^{k+2}h(z) = g(z), \qquad \text{Re } z > b.$$

Since this is true whenever $b > \max \{a, 0\}$, the proof is complete in the case $a \geq 0$.

When $a < 0$, let

$$g_1(z) = g(z + a).$$

Then g_1 is holomorphic for Re $z > 0$ and satisfies

$$|g_1(z)| \leq K_2(1 + |z|)^k \exp(-M \text{ Re } z).$$

It follows that $g_1 = LF_1$ for Re $z > 0$, some $F_1 \in \mathscr{L}'$. Then

$$g = LF, \qquad F = e_{-a}F_1. \qquad\qquad \square$$

Exercises

1. Compute the Laplace transforms of $D^k\delta$, $k = 0, 1, 2, \ldots$ and of $T_s\delta$, $s \in \mathbb{R}$.

2. Compute the Laplace transform of F when

$$F(u) = \int_0^\infty e^{ut}u(t) \, dt.$$

§7. Differential equations

In §§5, 6 of Chapter 2 we discussed differential equations of the form

$$u'(x) + au(x) = f(x),$$

and of the form

$$u''(x) + bu'(x) + cu(x) = f(x).$$

In this section we turn to the theory and practice of solving general nth order linear differential equations with constant coefficients:

(1) $$a_nu^{(n)} + a_{n-1}u^{(n-1)} + \cdots a_1u' + a_0u = f,$$

where the a_j are complex constants. Using D to denote differentiation, and understanding D^0 to be the identity operator, $D^0 u = u$, we may write (1) in the form

(1)′
$$\sum_{k=0}^{n} a_k D^k u = f.$$

Let p be the polynomial

$$p(z) = \sum_{k=0}^{n} a_k z^k.$$

Then it is natural to denote by $p(D)$ the *operator*

(2)
$$p(D) = \sum_{k=0}^{n} a_k D^k.$$

Equation (1) becomes

(1)″
$$p(D)u = f.$$

We shall assume that the polynomial is actually of degree n, that is

$$a_n \neq 0.$$

Before discussing (1)″ for *functions*, let us look at the corresponding problem for *distributions*: given $H \in \mathscr{L}'$, find $F \in \mathscr{L}'$ such that

$$p(D)F = H.$$

Theorem 7.1. *Suppose p is a polynomial of degree $n > 0$, and suppose $H \in \mathscr{L}'$. Then there is a unique distribution $F \in \mathscr{L}'$ such that*

(3)
$$p(D)F = H.$$

Proof. Distributions in $\mathscr{L}'$ are uniquely determined by their Laplace transforms. Therefore (3) is equivalent to

(4)
$$L(p(D)F)(z) = LH(z), \qquad \operatorname{Re} z > a$$

for some $a \in \mathbb{R}$. But

$$L(p(D)F)(z) = p(z)LF(z).$$

We may choose a so large that $p(z) \neq 0$ if $\operatorname{Re} z \geq a$, and so that LH is holomorphic for $\operatorname{Re} z > a$ and satisfies the estimate given in Theorem 6.3. Then we may define

$$g(z) = p(z)^{-1}LH(z), \qquad \operatorname{Re} z > a.$$

Then g is holomorphic, and it too satisfies estimates

$$|g(z)| \leq K(1 + |z|)^k \exp(-M \operatorname{Re} z), \qquad \operatorname{Re} z > a.$$

Theorem 6.3 assures us that there is a unique $F \in \mathscr{L}'$ such that $LF = g$, and then (4) holds. □

The proof just given provides us, in principle, with a way to calculate F, given H. Let us carry out the calculation formally, treating F and H as though they were functions:

$$F(t) = \frac{1}{2\pi i} \int_C e^{tz} LF(z) \, dz$$

$$= \frac{1}{2\pi i} \int_C e^{tz} p(z)^{-1} LH(z) \, dz$$

$$= \frac{1}{2\pi i} \int_C \int_{\mathbb{R}} e^{tz} p(z)^{-1} e^{-sz} H(s) \, ds \, dz$$

$$= \int_{\mathbb{R}} \left[\frac{1}{2\pi i} \int_C e^{(t-s)z} p(z)^{-1} \, dz \right] H(s) \, ds$$

$$= \int_{\mathbb{R}} G(t - s) H(s) \, ds,$$

where

(5)
$$G(t) = \frac{1}{2\pi i} \int_C e^{tz} p(z)^{-1} \, dz$$

and C is a line $\operatorname{Re} z = b > a$.

We emphasize that the calculation was purely formal. Nevertheless the integral (5) makes sense if p has degree ≥ 2, and defines a function G. Equivalently,

(5)′
$$G(t) = \lim_{R \to \infty} \frac{1}{2\pi i} \int_{C_R} e^{tz} p(z)^{-1} \, dz,$$

where C_R is the directed line segment from $b - iR$ to $b + iR$, $R > 0$. We shall show that the limit (5)′ also exists when p has degree 1, except when $t = 0$. The function defined by (5)′ is called the *Green's function* for the operator $p(D)$ defined by the polynomial p. Our formal calculation suggests that G plays a central role in solving differential equations. The following two theorems provide some information about it.

Theorem 7.2. *Suppose p is a polynomial of degree $n \geq 1$; suppose $z_1, z_2, \ldots, z_r$ are the distinct roots of p, and suppose that z_j has multiplicity m_j. Then (5)′ defines a function G for all $t \neq 0$. This function is a linear combination of the functions g_{jk}, $1 \leq j \leq r$, $0 \leq k < m_j$, where*

$$g_{jk}(t) = 0, \qquad t < 0;$$
$$g_{jk}(t) = t^k \exp(z_j t), \qquad t > 0.$$

Proof. Suppose $t < 0$. Let D_R denote the rectangle with vertices $b \pm iR$, $(b + R^{1/2}) \pm iR$. When $\operatorname{Re} z \geq b$,

(6)
$$|e^{zt} p(z)^{-1}| \leq c(t)(1 + |z|)^{-n} \exp t(\operatorname{Re} z - b),$$

where $c(t)$ is independent of z. Now the line segment C_R is one side of the rectangle D_R, and the estimates (6) show that the integral of $e^{zt}p(z)^{-1}$ over the other sides converges to 0 as $R \to \infty$. On the other hand, the integral over all of D_R vanishes, because the integrand is holomorphic inside D_R. Thus the limit in (5)′ exists and is 0 when $t < 0$ (and also when $t = 0$, if $n > 1$).

Suppose $t > 0$. Let D_R now be the rectangle with vertices $b \pm iR$, $(b - R^{1/2}) \pm iR$, with the counterclockwise direction, and suppose R is so large that D_R contains all roots of $p(z)$. Then the integral of $e^{zt}p(z)^{-1}$ over D_R is independent of R, and the integral over the sides other than C_R tends to 0 as $R \to \infty$. Thus again the limit in (5)′ exists, and

$$(5)'' \qquad\qquad G(t) = \frac{1}{2\pi i} \int_{D_R} e^{zt}p(z)^{-1}\,dz, \qquad t > 0.$$

Now we may apply Theorem 6.3 of Chapter 6: $G(t)$ is the sum of the residues of the meromorphic function $e^{zt}p(z)^{-1}$. The point z_j is a pole of order m_j, so near z_j we have a Laurent expansion

$$p(z)^{-1} = \sum_{m \ge -m_j} b_m(z - z_j)^m.$$

Combining this with

$$e^{zt} = e^{z_j t} \sum_{m \ge 0} (m!)^{-1}(z - z_j)^m t^m,$$

we see that the residue (the coefficient of $(z - z_j)^{-1}$ in the Laurent expansion) at z_j is a linear combination of

$$t^k \exp(z_j t), \qquad 0 \le k < m_j;$$

moreover, it is the same linear combination whatever the value of t. $\square$

Suppose f is a complex-valued function defined on an interval (a, b). We write

$$f(a+) = \lim_{t \to a} f(a)$$

when the limit on the right exists as t approaches from the right.

We take the Green's function for $p(D)$ to be 0 at $t = 0$; when $n > 1$ this agrees with (5)′.

Theorem 7.3. *Let p be a polynomial of degree $n > 0$, with leading coefficient $a_n \ne 0$. Let G be the Green's function for $p(D)$. Then G is the unique function from $\mathbb{R}$ to $\mathbb{C}$ having the following properties:*

(7)   *all derivatives $D^k G$ exist and are continuous when $t \ne 0$;*

(8)   *the derivatives $D^k G$ exist and are continuous at 0 when $k \le n - 2$;*

$$(9) \qquad\qquad\qquad G(t) = 0, \qquad t \le 0;$$

$$(10) \qquad\qquad\qquad p(D)G(t) = 0, \qquad t > 0;$$

$$(11) \qquad\qquad\qquad a_n D^{n-1}G(0+) = 1.$$

Proof. We know that G is a linear combination of functions satisfying (7) and (9), so G does also. When $t > 0$ we may differentiate (5)″ and get

(12)
$$D^k G(t) = \frac{1}{2\pi i} \int_{D_R} z^k e^{zt} p(z)^{-1} \, dz.$$

Thus

$$D^k G(0+) = \frac{1}{2\pi i} \int_{D_R} z^k p(z)^{-1} \, dz.$$

We may replace D_R by a very large circle centered at the origin and conclude that

$$D^k G(0+) = 0, \qquad k \le n - 2.$$

Therefore (8) is true. Let us apply the same argument when $k = n - 1$. Over the large circle the integrand is close to

$$z^{n-1}(a_n z_n)^{-1} = a_n^{-1} z^{-1},$$

so

$$D^{n-1} G(0+) = a_n^{-1}.$$

Finally, (12) gives

$$p(D)G(t) = \frac{1}{2\pi i} \int_{D_R} e^{zt} \, dz = 0, \qquad t > 0.$$

Now we must show that G is uniquely determined by the properties (7)–(11). Suppose G_1 also satisfies (7)–(11), and let $f = G - G_1$. Then f satisfies (7)–(10); moreover $D^{n-1} f(0) = 0$. We may factor

$$p(z) = a_n(z - z_1)(z - z_2) \cdots (z - z_n),$$

where we do not assume that the z_j are distinct. Let $f_0 = f$, and let

$$f_k = D f_{k-1} - z_k f_{k-1}, \qquad k > 0.$$

Then each f_k is a linear combination of $D^j f$, $0 \le j \le k$, so

$$f_k(0) = 0, \qquad k \le n - 1.$$

Moreover,

$$\begin{aligned}
f_n &= (D - z_n)(D - z_{n-1}) \cdots (D - z_1) f \\
&= a_n^{-1} p(D) f = 0.
\end{aligned}$$

Thus

$$f_{n-1}(0) = 0, \qquad D f_{n-1} - z_n f_{n-1} = f_n = 0,$$

so

$$f_{n-1} = 0.$$

Then

$$f_{n-2}(0) = 0, \qquad D f_{n-2} - z_{n-1} f_{n-2} = 0,$$

so $f_{n-2} = 0$. (We are using Theorem 5.1 of Chapter 2). Inductively, each $f_k = 0$, $k \le n$, so $f = 0$ and $G = G_1$. $\square$

Now let us return to differential equations for functions.

Theorem 7.4. *Suppose p is a polynomial of degree $n > 0$, and suppose $f: [0, \infty) \to \mathbb{C}$ is a continuous function. Then there is a unique solution $u: [0, \infty) \to \mathbb{C}$ to the problem*

(13) $p(D)u(t) = f(t)$, $t > 0$;

(14) $D^j u(0+) = 0$, $0 \le j \le n - 1$.

This solution u is given by

(15) $$u(t) = \int_0^t G(t - s)f(s)\, ds,$$

where G is the Green's function for the operator $p(D)$.

Proof. We use properties (7)–(11) of G. Let u be given by (15) for $t \ge 0$. Then successive differentiations yield

(16) $$Du(t) = G(0+)f(t) + \int_0^t DG(t - s)f(s)\, ds$$

$$= \int_0^t DG(t - s)f(s)\, ds, \ldots,$$

(17) $$D^k u(t) = \int_0^t D^k G(t - s)f(s)\, ds, \qquad k \le n - 1,$$

(18) $$D^n u(t) = a_n^{-1} f(t) + \int_0^t D^n G(t - s)f(s)\, ds.$$

Thus

$$p(D)u(t) = f(t) + \int_0^t p(D)G(t - s)f(s)\, ds$$

$$= f(t).$$

Moreover, (17) implies (14). Thus u is a solution. The uniqueness of u is proved in the same way as uniqueness of G. $\square$

We conclude with a number of remarks.

1. The problem (13)–(14) as a problem for distributions: Let us define $f(t) = 0$ for $t < 0$. If f does not grow too fast, i.e., if for some $a \in \mathbb{R}$

$$e^{-at} f(t) \quad \text{is bounded,}$$

then we may define a distribution $H \in \mathscr{L}'$ by

$$H(v) = \int f(t)v(t)\, dt, \qquad v \in \mathscr{L}.$$

Suppose u is the solution of (13)–(14). Then it can be shown that u defines a distribution F, and

(19) $$p(D)F = H.$$

Thus we have returned to the case of Theorem 7.1.

2. If the problem (13)–(14) is reduced to (19), then the proof of Theorem 7.1 shows that the solution may be found by determining its Laplace transform. Since there are extensive tables of Laplace transforms, this is of practical as well as theoretical interest. It should be noted that Laplace transform tables list functions which are considered to be defined only for $t \geq 0$; then the Laplace transform of such a function f is taken to be

$$Lf(z) = \int_0^\infty e^{-zt} f(t)\, dt.$$

In the context of this chapter, this amounts to setting $f(t) = 0$ for $t < 0$ and considering the distribution determined by f, exactly as in Remark 1.

3. Let us consider an example of the situation described in Remark 2. A table of Laplace transforms may read, in part,

f	Lf
$\sin t$	$(z^2 + 1)^{-1}$
$\sinh t$	$(z^2 - 1)^{-1}$

(As noted in Remark 2, the function $\sin t$ in the table is considered only for $t \geq 0$, or is extended to vanish for $t < 0$.)

Now suppose we wish to solve:

(20) $$u''(t) - u(t) - \sin t = 0, \qquad t > 0;$$

(21) $$u(0) = u'(0) = 0.$$

Let $p(z) = z^2 - 1$. Our problem is

$$p(D)u = \sin t, \qquad t > 0;$$
$$u(0) = Du(0) = 0.$$

The solution u is the function whose Laplace transform is

$$p(z)^{-1} L(\sin t)(z) = (z^2 - 1)^{-1}(z^2 + 1)^{-1}.$$

But

$$(z^2 - 1)^{-1}(z^2 + 1)^{-1} = \tfrac{1}{2}(z^2 - 1)^{-1} - \tfrac{1}{2}(z^2 + 1)^{-1}.$$

Therefore the solution to (20)–(21) is

$$u(t) = \tfrac{1}{2} \sinh t - \tfrac{1}{2} \sin t, \qquad t \geq 0.$$

4. In cases where the above method fails, either because the given function f grows too fast to have a Laplace transform or because the function Lu cannot be located in a table, one may wish to compute the Green's function G and use (15). The Green's function may be computed explicitly if the roots of the polynomial p are known (of course (5)″ gives us G *in principle*). In fact, suppose the roots are $z_1, z_2, \ldots, z_r$ with multiplicities $m_1, m_2, \ldots, m_r$. We know that G for $t > 0$, is a linear combination of the n functions

$$t^k \exp(z_j t), \qquad k < m_j.$$

Thus

$$G(t) = \sum c_{jk} t^k \exp(z_j t), \qquad t > 0,$$

where we must determine the constants c_{jk}. The conditions (8) and (11) give n independent linear equations for these n constants. In fact,

$$G(0+) = \sum c_{j0},$$
$$DG(0+) = \sum z_j c_{j0} + \sum c_{j1},$$

etc.

5. The more general problem

(22) $$\qquad\qquad p(D)u(t) = f(t), \qquad t > 0;$$

(23) $$\qquad\qquad D^k u(0+) = b_k, \qquad 0 \le k < n$$

may be reduced to (13)–(14). Two ways of doing this are given in the exercises.

6. The formal calculation after Theorem 7.1 led to a formula

$$F(t) = \int G(t - s) H(s) \, ds$$

which it is natural to interpret as a convolution (see Chapter 3). A brief sketch of such a development is given in the exercises.

Exercises

1. Compute the Green's function for the operator $p(D)$ in each of the following cases:

$$p(z) = z^2 - 4z - 5$$
$$p(z) = z^2 - 4z + 4$$
$$p(z) = z^3 + 2z^2 - z - 2$$
$$p(z) = z^3 - 3z + 2.$$

2. Solve for u:

$$u''(t) - 4u'(t) + 4u(t) = e^t, \qquad t > 0,$$
$$u(0+) = u'(0+) = 0.$$

3. Solve for u:

$$u'''(t) - 3u'(t) + 2u(t) = t^2 - \cos t, \qquad t > 0,$$
$$u(0+) = u'(0+) = u''(0+) = 0$$

4. Let $u_0: (0, \infty) \to \mathbb{C}$ be given by

$$u_0(t) = \sum_{k=0}^{n-1} (k!)^{-1} b_k t^k.$$

Show that

$$D^k u_0(0+) = b_k, \qquad 0 \le k \le n - 1.$$

5. Suppose $u_0: (0, \infty) \to \mathbb{R}$ is such that

$$D^k u_0(0+) = b_k, \qquad 0 \le k \le n - 1.$$

Show that u is a solution of (22)–(23) if and only if $u = u_0 + u_1$, where u_1 is the solution of

$$p(D)u_1(t) = f(t) - p(D)u_0(t), \qquad t > 0,$$
$$D^k u_1(0+) = 0, \qquad 0 \le k \le n - 1.$$

6. Show that problem (22)–(23) has a unique solution.
7. Show that the solution of

$$p(D)u(t) = 0, \qquad t > 0,$$
$$D^k u(0+) = 0, \qquad 0 \le k < j \quad \text{and} \quad j < k \le n - 1,$$
$$D^j u(0+) = 1$$

is a linear combination of functions $t^k \exp zt$.

8. Show that any solution of

$$p(D)u(t) = 0, \qquad t > 0$$

is a linear combination of the functions $t^k \exp zt$, where z is a root of $p(D)$ with multiplicity greater than k, and conversely.

9. Suppose $u: (0, \infty) \to \mathbb{C}$ is smooth and suppose $D^k u(0+)$ exists for each k. Suppose also that each $D^k u$ defines a distribution F_k by

$$F_k(v) = \int_0^\infty D^k u(t) v(t) \, dt.$$

Show that

$$DF_0 = F_1 + u(0+)\delta,$$

and in general

$$D^k F_0 = F_k + \sum_{j=0}^{k-1} D^j u(0+) D^{k-1-j} \delta.$$

10. In Exercise 9 let $u(t) = G(t)$, $t > 0$, where G is the Green's function for $p(D)$. Show that

$$p(D)F_0 = \delta.$$

11. Use Exercise 9 to interpret the problem (22)–(23) as a problem of finding a distribution (when the function f defines a distribution in $\mathscr{L}'$). Discuss the solution of the problem.

12. Use Exercise 11 to give another derivation of the result of Exercise 5.

13. Again let

$$\tilde{u}(t) = u(-t), \qquad T_s u(t) = u(t-s).$$

If $F \in \mathscr{L}'$ and $u \in \mathscr{L}$, set

$$\tilde{F} * u(t) = F(T_{-t}u).$$

(a) Suppose $F = F_v$, where $v \colon \mathbb{R} \to \mathbb{C}$ is continuous, $v(t) = 0$ for $t \le -M$, and $e^{-at}v(t)$ is bounded. Show that for each $u \in \mathscr{L}$ the convolution integral

$$\tilde{v} * u(t) = \int \tilde{v}(t-s)u(s)\, ds$$

exists and equals

$$\tilde{F} * u(t).$$

(b) Show that for each $F \in \mathscr{L}'$ and $u \in \mathscr{L}$, the function $\tilde{F} * u$ is in $\mathscr{L}$.

14. If $F, H \in \mathscr{L}'$, set

$$(F * H)(u) = F(\tilde{H} * u), \qquad u \in \mathscr{L}.$$

Show that $F * H \in \mathscr{L}'$.

15. Compute $(D^k \delta)^{\sim} * u$, $u \in \mathscr{L}$. Compute $(D^k \delta) * F$, $F \in \mathscr{L}'$.

16. Show that

$$L(F * H) = L(F)L(H).$$

17. Let G be the distribution determined by the Green's function for $p(D)$. Show that

$$LG = p(z)^{-1}.$$

18. Show that the solution of

$$p(D)F = H$$

is

$$F = G * H,$$

where G is as in Exercise 17.

NOTES AND BIBLIOGRAPHY

Chapters 1 and 2. The book by Kaplansky [9] is a very readable source of further material on set theory and metric spaces. The classical book by Whittaker and Watson [25] and the more modern one by Rudin [18] treat the real and complex number systems, compactness and continuity, and the topics of Chapter 2. Vector spaces, linear functionals, and linear transformations are the subject of any linear algebra text, such as Halmos [7]. Infinite sequences and series may be pursued further in the books of Knopp [10], [11]. More problems (and theorems) in analysis are to be found in the classic by Polya and Szegő [15].

Chapters 3, 4, and 5. The Weierstrass theorems (and the technique of approximation by convolution with an approximate identity) are classical. A direct proof of the polynomial approximation theorem and a statement and proof of Stone's generalization may be found in Rudin [18].

The general theory of distributions (or "generalized functions") is due to Laurent Schwartz, and is expounded in his book [20]. The little book by Lighthill [12] discusses periodic distributions and Fourier series. Other references for distribution theory and applications are the books of Bremermann [2], Liverman [13], Schwartz [21], and Zemanian [27].

Banach spaces, Frechet spaces, and generalizations are treated in books on functional analysis: that by Yosida [26] is comprehensive; the treatise by Dunford and Schwartz [4] is exhaustive; the sprightly text by Reed and Simon [16] is oriented toward mathematical physics. Good sources for Hilbert space theory in particular are the books by Halmos [6], [8] and by Riesz and Sz.-Nagy [17].

The classical L^2-theory of Fourier series treats $L^2(0, 2\pi)$ as a space of functions rather than as a space of distributions, and requires Lebesgue integration. Chapters 11 through 13 of Titchmarsh [24] contain a concise development of Lebesgue integration and the L^2-theory. A more leisurely account is in Sz.-Nagy [14]. The treatise by Zygmund [28] is comprehensive.

Chapter 6. The material in §1–§6 is standard. The classic text by Titchmarsh [24] and that by Ahlfors [1] are good general sources. The book by Rudin [19] also treats the boundary behavior of functions in the disc, related to the material in §7.

Chapter 7. The Laplace transform is the principal subject of most books on "operational mathematics" and "transform methods." Doetsch [3] is a comprehensive classical treatise. Distribution-theoretic points of view are presented in the books of Bremermann [2], Erdelyi [5], Liverman [13], and Schwartz [21].

Bibliography

1. AHLFORS, L. V.: *Complex Analysis*, 2nd ed. New York: McGraw-Hill 1966.
2. BREMERMANN, H. J.: *Distributions, Complex Variables, and Fourier Transforms*. Reading, Mass.: Addison-Wesley 1966.

3. DOETSCH, G.: *Handbuch der Laplace-Transformation*, 3 vols. Basel: Birkhauser 1950–1956.

5. ERDELYI, A.: *Operational calculus and generalized functions.* New York: Holt, Rinehart & Winston 1962.

4. DUNFORD, N., and SCHWARTZ, J. T.: *Linear Operators*, 3 Vols. New York: Wiley-Interscience 1958–1971.

6. HALMOS, P. R.: *Introduction to Hilbert Space*, 2nd. ed. New York: Chelsea 1957.

7. HALMOS, P. R.: *Finite Dimensional Vector Spaces.* Princeton: Van Nostrand 1958.

8. HALMOS, P. R.: *A Hilbert Space Problem Book.* Princeton: Van Nostrand 1967.

9. KAPLANSKY, I.: *Set Theory and Metric Spaces.* Boston: Allyn & Bacon 1972.

10. KNOPP, K.: *Theory and Applications of Infinite Series.* London and Glasgow: Blackie & Son 1928.

11. KNOPP, K.: *Infinite Sequences and Series.* New York: Dover 1956.

12. LIGHTHILL, M. J.: *Introduction to Fourier Analysis and Generalized Functions.* New York: Cambridge University Press 1958.

13. LIVERMAN, T. P. G.: *Generalized Functions and Direct Operational Methods.* Englewood Cliffs, N.J.: Prentice-Hall 1964.

14. SZ.-NAGY, B.: *Introduction to Real Functions and Orthogonal Expansions.* New York: Oxford University Press 1965.

15. POLYA, G., and SZEGŐ, G.: *Problems and Theorems in Analysis.* Berlin–Heidelberg–New York: Springer 1972.

16. REED. M., and SIMON, B.: *Methods of Modern Mathematical Physics*, vol. 1. New York: Academic Press 1972.

17. RIESZ, F., and SZ.-NAGY, B.: *Functional Analysis.* New York: Ungar 1955.

18. RUDIN, W.: *Principles of Mathematical Analysis*, 2nd ed. New York: McGraw-Hill 1964.

19. RUDIN, W.: *Real and Complex Analysis.* New York: McGraw-Hill 1966.

20. SCHWARTZ, L.: *Theorie des Distributions*, 2nd ed. Paris: Hermann 1966.

21. SCHWARTZ, L.: *Mathematics for the Physical Sciences.* Paris: Hermann; Reading, Mass.: Addison-Wesley 1966.

22. SOBOLEV, S. L.: *Applications of Functional Analysis in Mathematical Physics.* Providence: Amer. Math. Soc. 1963.

23. SOBOLEV, S. L.: *Partial Differential Equations of Mathematical Physics.* Oxford and New York: Pergamon 1964.

24. TITCHMARSH, E. C.: *The Theory of Functions*, 2nd ed. London: Oxford University Press 1939.

25. WHITTAKER, E. T., and WATSON, G. N.: *A Course of Modern Analysis*, 4th ed. London: Cambridge University Press 1969.

26. YOSIDA, K.: *Functional Analysis*, 2nd ed. Berlin-Heidelberg-New York: Springer 1968.

27. ZEMANIAN, A. H.: *Distribution Theory and Transform Analysis.* New York: McGraw-Hill 1965.

28. ZYGMUND, A.: *Trigonometric Series*, 2nd ed. London: Cambridge University Press 1968.

NOTATION INDEX

SUBJECT INDEX

Graduate Texts in Mathematics

continued from page ii